Community, Immunity and the Proper

It is widely apparent in our hyper-globalized world that the epistemologies, institutions and practices underwriting it have reached a state of profound crisis. In the globalized world, everything is inevitably brought into proximity and correlation. Wars, natural disasters, climatic upheaval, political and economic turmoil – none of these can be effectively isolated, insulated, instituted, even immunized, as something apart, something that might be considered proper only to itself. This collected edition considers this crisis of the proper with a focus on Italian political theorist Roberto Esposito's work on community, immunity and biopolitics.

This collection introduces Esposito's work to a wider English-speaking audience and provides many important contributions to the burgeoning scholarship on his political theory. Important international scholars working in this area examine and analyze his theory from a variety of perspectives, including those of biopolitics, feminism, political theory, the history of philosophy (Spinoza, Hegel, Heidegger and Jean-Luc Nancy), property, community and gift economies. The collection also includes previously untranslated essays by Esposito and Jean-Luc Nancy. This collection will be of interest to those just discovering Esposito and for those who are already familiar with his work.

This book was originally published as a Special Issue of *Angelaki: Journal of the Theoretical Humanities*.

Greg Bird is an Assistant Professor in the Department of Sociology at Wilfrid Laurier University, Canada. His research examines critical iterations of community in classical and contemporary social and political philosophy.

Jonathan Short teaches in the Department of Sociology at York University in Toronto, Canada. He has published essays in contemporary social and political thought with an emphasis on contemporary Italian political theory.

Community, Immunity and the Proper

Roberto Esposito

Edited by
Greg Bird and Jonathan Short

Routledge
Taylor & Francis Group
LONDON AND NEW YORK

First published 2015 by Routledge

2 Park Square, Milton Park, Abingdon, Oxon OX14 4RN
711 Third Avenue, New York, NY 10017, USA

First issued in paperback 2017

Routledge is an imprint of the Taylor & Francis Group, an informa business

British Library Cataloguing in Publication Data
A catalogue record for this book is available from the British Library

ISBN 13: 978-1-138-84895-5 (hbk)
ISBN 13: 978-1-138-05748-7 (pbk)

Typeset in Bodoni MT
by RefineCatch Limited, Bungay, Suffolk

Publisher's Note
The publisher accepts responsibility for any inconsistencies that may have arisen during the conversion of this book from journal articles to book chapters, namely the possible inclusion of journal terminology.

Disclaimer
Every effort has been made to contact copyright holders for their permission to reprint material in this book. The publishers would be grateful to hear from any copyright holder who is not here acknowledged and will undertake to rectify any errors or omissions in future editions of this book.

Contents

Citation Information

The chapters in this book were originally published in *Angelaki: Journal of the Theoretical Humanities*, volume 18, issue 3 (September 2013). When citing this material, please use the original page numbering for each article, as follows:

Chapter 1
Community, Immunity, and the Proper: An Introduction to the Political Theory of Roberto Esposito
Greg Bird and Jonathan Short
Angelaki, volume 18, issue 3 (September 2013) pp. 1–12

Chapter 2
Hegel on Communitas*: An Unexplored Relationship between Hegel and Esposito*
María del Rosario Acosta López
Angelaki, volume 18, issue 3 (September 2013) pp. 13–32

Chapter 3
Roberto Esposito's Deontological Communal Contract
Greg Bird
Angelaki, volume 18, issue 3 (September 2013) pp. 33–48

Chapter 4
The Membrane and the Diaphragm: Derrida and Esposito on Immunity, Community, and Birth
Penelope Deutscher
Angelaki, volume 18, issue 3 (September 2013) pp. 49–68

Chapter 5
How (Not) to Properly Abandon the Improper?
Ignaas Devisch
Angelaki, volume 18, issue 3 (September 2013) pp. 69–82

Chapter 6
Community, Immunity, Biopolitics
Roberto Esposito (translated by Zakiya Hanafi)
Angelaki, volume 18, issue 3 (September 2013) pp. 83–90

Chapter 7
Spinoza and the Biopolitical Roots of Modernity
Peter Gratton
Angelaki, volume 18, issue 3 (September 2013) pp. 91–102

Chapter 8
The Ethics of Community: Nancy, Blanchot, Esposito
Kristin Hole
Angelaki, volume 18, issue 3 (September 2013) pp. 103–118

Chapter 9
Fraternity
Jean-Luc Nancy (translated by Sarah Clift)
Angelaki, volume 18, issue 3 (September 2013) pp. 119–124

Chapter 10
Communitas *and the Problem of Women*
Anne O'Byrne
Angelaki, volume 18, issue 3 (September 2013) pp. 125–138

Chapter 11
On an Obligatory Nothing: Situating the Political in Post-Metaphysical Community
Jonathan Short
Angelaki, volume 18, issue 3 (September 2013) pp. 139–154

Chapter 12
Roberto Esposito's Political Philosophy of the Gift
Lorna Weir
Angelaki, volume 18, issue 3 (September 2013) pp. 155–168

Please direct any queries you may have about the citations to
clsuk.permissions@cengage.com

Notes on Contributors

María del Rosario Acosta López is an Associate Professor in the Department of Philosophy at DePaul University, Chicago, USA, and formerly a guest researcher in the "Normative Orders" Excellence Cluster, Goethe Universitat Frankfurt, funded by the Alexander von Humboldt Foundation. She has compiled books on Hegel, Schiller, contemporary philosophy of art and contemporary political philosophy. Some of her latest published papers explore the relationship between Hegel's political philosophy and the notion of "community" within a certain contemporary French tradition (Derrida, Esposito, Nancy and Agamben). Her current research project is a book on Hegel from the standpoint of said debate.

Greg Bird is an Assistant Professor in the Department of Sociology at Wilfrid Laurier University, Waterloo, Canada. His research examines critical iterations of community in classical and contemporary social and political theory. Besides publications on phenomenology, Esposito and Nancy, he has a forthcoming publication in Italian on the dialectic of alienation and appropriation in revolutionary thought. In 2014, he acted as a visiting fellow at the Istituto Italiano di Scienze Umane in Naples, Italy, where he was working on his book *Containing Community*, which examines how the proper is formulated in the debate about community between Agamben, Esposito and Nancy (under review). He is currently working on a series of articles that examine the relationship between immunization and community in temporary foreign worker programs and in contemporary biopolitical formulations in Italian Theory.

Sarah Clift is an Assistant Professor of Contemporary Studies at the University of King's College, Halifax, Canada. She is the author of *Committing the Future to Memory: History, Memory, Trauma* (2013), and is active as a translator of German and French philosophy. Her translations include Jean Luc-Nancy's *God, Justice, Love, Beauty: Four Little Dialogues* (2011), Aleida Assmann's *The Long Shadow of the Past*, and, with Pascale-Anne Brault and Michael Naas, Jean- Luc Nancy's *Noli me Tangere: On the Raising of the Body* (2008).

Penelope Deutscher is a Professor of Philosophy at Northwestern University, Evanston, Illinois, USA. Her publications include *Yielding Gender: Feminism, Deconstruction and the History of Philosophy* (1997), *A Politics of Impossible Difference: The Later Work of Luce Irigaray* (2002), *How to Read Derrida* (2006), and *The Philosophy of Simone de Beauvoir: Conversion, Ambiguity, Resistance* (2008). Essays from her current project on reproduction and post-Foucauldian biopolitics have been published in *South Atlantic Quarterly, Journal of Bioethical Inquiry, Theory, Culture and Society* and *Telos*.

Ignaas Devisch is a Professor in Ethics, Philosophy and Medical Philosophy. He holds positions at Ghent University, Belgium and University College Arteveldehogeschool,

Belgium. He publishes in the fields of social and political Continental philosophy as well as medical philosophy, philosophy of sports and ethics. Recent volumes include his *Jean-Luc Nancy and the Question of Community* (2012) and his co-authored *Re-treating Religion* (2012).

Roberto Esposito is Professor of Theoretical Philosophy at Scuola Normale Superiore di Pisa. He has lectured and given seminars at major universities around the world, including Harvard, UCLA, Columbia, Brown, Duke, Michigan and Cornell in the United States; and in England, France, Germany, Spain, Portugal, Austria, Japan, Argentina and Chile. He was one of the founders of the European Political Lexicon Research Centre and the International Centre for a European Legal and Political Lexicon. For five years he was the only Italian member of the International Council of Scholars of the Collège International de Philosophie (CIP) in Paris. Currently he co-edits the review *Filosofia Politica* published by il Mulino, the series *Comunità e Libertà* for Laterza, and acts as a philosophy consultant for the publisher Einaudi. His books are translated into English, French, German, Spanish, Portuguese, Japanese and Korean.

Peter Gratton is in the Department of Philosophy at the Memorial University of Newfoundland, Canada. He has published numerous articles in political, Continental, and intercultural philosophy and is the author of *The State of Sovereignty: Lessons from the Political Fictions of Modernity* (2012) and *Speculative Realism* (forthcoming). He is co-editor of the influential interdisciplinary journal *Society and Space*, books editor of *Derrida Today*, and board member of national philosophical societies. Peter has also edited two works: *Traversing the Imaginary* (2007), co-edited with John Mannousakis and Jean-Luc Nancy, and *Plural Thinking: Expositions of World, Politics, Art, and Sense* (2012), co-edited with Marie-Eve Morin.

Zakiya Hanafi received her Ph.D. in Italian Studies from Stanford University, USA, in 1991. She is the author of *The Monster in the Machine: Magic, Medicine, and the Marvellous in the Time of the Scientific Revolution* (2000) as well as the translator of numerous works of Italian and French philosophy, including Roberto Esposito's *Immunitas* (2011), *Third Person* (2012), and *Living Thought* (2012). Formerly an Adjunct Professor at the University of Venice (Ca' Foscari), Italy, she currently lectures at the University of Washington, Seattle, USA.

Kristin Hole holds a PhD in Comparative Literary and Cultural Studies from Stony Brook University. She has published in *Angelaki* and in the edited volume *Fashion and Film* (Indiana UP, 2011). Her forthcoming book is entitled *Towards a Feminist Cinematic Ethics: Claire Denis, Emmanuel Levinas, and Jean-Luc Nancy* (Edinburgh University Press, 2015). She teaches in the Film Studies program at Portland State University.

Jean-Luc Nancy is a Professor of Philosophy at the University of Strasbourg, France, and a member of the faculty at the European Graduate School, Saas-Fee, Switzerland. He is one of the leading figures in contemporary French thought. He has published more than fifty books, most of which have been translated into English. Recent translations include *Corpus*, *The Ground of the Image*, *Listening*, and *Disenclosure: The Deconstruction of Christianity*.

Anne O'Byrne is an Associate Professor of Philosophy at Stony Brook University, New York, USA. Her work is in political philosophy and ontology, and engages a number of thinkers in twentieth-century and contemporary Continental philosophy – Heidegger, Arendt, Dilthey, Derrida and Jean-Luc Nancy – on a range of issues, namely identity, natality, embodiment, education, history, gender and genocide. Publications include *Natality and Finitude* (2010), several articles, and translations of Nancy's *Being Singular Plural* (translated with Robert Richardson, 1996) and *Corpus II* (2013).

Jonathan Short is presently an adjunct faculty member in the Department of Sociology at York University, Toronto, Canada. In 2012 he was a researcher in the Theory Department at the Jan Van Eyck Academy in Maastricht, The Netherlands. He has published essays in contemporary social and political thought with an emphasis on contemporary Italian political theory, and is writing a series of papers on the relationship between community, legality and the political.

Lorna Weir is a Professor of Sociology and Social and Political Thought at York University, Toronto, Canada. Her books include *Pregnancy, Risk and Biopolitics: On the Threshold of the Living Subject* (Routledge, 2006) and *Global Public Health Vigilance: Creating a World on Alert* (Routledge, 2010, with Eric Mykhalovskiy). She has published widely on biopolitics, including its relation to the theory of sovereignty, in the *European Journal of Social Theory, Thesis 11, Feminist Studies, Social Science and Medicine* and *Systems and Synthetic Biology*. Her current research investigates diplomatic gifts in our present, the genealogy of biosecurity and sacrificial processes in contemporary health systems.

greg bird
jonathan short

COMMUNITY, IMMUNITY, AND THE PROPER

an introduction to the political theory of roberto esposito

introduction

It is widely apparent in our hyper-globalized world that the epistemologies, institutions, and practices underwriting it have reached a state of profound crisis. Intimately bound up with this sense of crisis is its apparent multiplicity and lack of isomorphism, as increasingly no single crisis can be seen to function independently of others. In the globalized world, everything is inevitably brought into proximity and correlation, be it wars, natural disasters, climatic upheaval, or political and economic turmoil. There is, accordingly, nothing that can be effectively isolated, insulated, instituted, even immunized, as something apart, something that might be considered *proper* only to itself. In this light the globalized world appears as the sustained crisis of the proper and simultaneously as the endgame of the project of modernization as manifested in ever more intensified, crisis-ridden forms. Even the very framework of crisis theory is itself starting to implode, thus becoming a crisis of second-order proportion.

The centrality of the concept of the proper to systemic crisis is similarly evident. For if the proper would indicate those items that are suitably, correctly, or even essentially joined together, it is precisely this obviousness of connection that is most troubled by the interlinked contemporary form of crisis. In this respect, if as Marx and Engels wrote more than a century ago, "all that is solid melts into air, all that is holy is profaned" to describe the globalizing dynamics of the capitalism of their day, they were simply identifying the crisis of the proper on which Western modernization was being built. The paradox of the proper is that the modality of crisis is essential to it. Internal to the associative matrix that links one element properly to another is that spacing Esposito inherits from Derrida and is further elaborated by Nancy – in turn radicalizing the analysis of temporization performed by Husserl – as the underlying structure of temporality in which what appears to be naturally related comes to be opened up, made subject to spacing, and thus exposed to an essential and necessary dissociation and dislocation.

Just as the proper implies an internally necessary connection to crisis, it is no exaggeration to say that politics today revolves around the dissociation and crisis of the proper. The more categorical the exposure of the proper to its essential contingency, the more violent are the political forms seeking to deny any such thing, to carry on with "business as usual" or to reassert the alleged naturalness of its associations in ever more apocalyptic terms. Today we are witnessing ever more drastic assertions of the essentiality of *propriety* (in the form of religious or socially conservative resurgence), *property* (in the form of highly concentrated but unstable regimes of capitalist accumulation), and/or *authenticity* (in the form of competing claims of idiomatic identity or questions of autonomous decision-making powers). Frantic attempts to reassert or bolster propriety, property, and authenticity, in short, the fields of the proper, are but a symptom of the latter's ineluctable crisis and decline.

It is amidst this background of uncertainty and the crisis of received forms that Roberto Esposito's political thought takes on an extreme relevance in the present. Esposito's work excavates and examines in a singularly powerful way how the dominant Western philosophical-political idealization of an immunized and proper community is becoming increasingly untenable in our current geo-political circumstances. Esposito's critique is not simply a nihilistic gesture, but is instead a prelude to a perhaps inevitable, but in any case long-overdue, rethinking of the basis of political and social relations. From his perspective, community is anything but a common essence or a shared property. The immunized models of community, where members are protected against foreign substances, external threats, and internal contagions, so common in our times, are imploding at a frightening pace. Rather than search for new material to mend the breaches of the communal borders and shield community against the nihilism of expropriation, Esposito searches for the original link between community and expropriation. Community does not shelter, contain, and protect us; rather, community is the very inauguration of an expropriation process. In community, so-called proprietary subjects are mutually exposed and suspended in a common *munus*, which never forms a stable property or furnishes an essential identity. The much-heralded crisis of community today, he contends, is merely the crisis of the project of community conceived under the banner of immunization. Community can no longer be conceived as an archi-original border that shelters the proper from being expropriated in its various senses. If the exigency of community is to be addressed in our times, then we are left with the seemingly impossible task of deconstructing the proper.[1]

This task has not been without its detractors. In the burgeoning English literature on Esposito, a single question is constantly being raised about his work: what kind of politics can come from such an approach? Most recognize that he is coming from somewhere on the left – his revision of community is grounded in a notion of communism – but his lack of concrete statements, his insistence that he is more committed to deconstructing core categories in modern political thought than prescribing practical political solutions, and his tendency to present his opinion while working through an interpretation of other theorists, have left many wondering if a concrete politics can be drawn out of his political theory.[2] This question appears to haunt Esposito himself, who has responded with slightly more practical statements in his recent *Terms of the Political* and in his article "Community, Immunity, Biopolitics" published in this special issue of *Angelaki*. The latter provides more fodder for those of us who are looking to situate his thought, but for those who want to see what they can do with his theory beyond applying it to concrete case studies – a strain of literature which is surely going to take off over the next decade – we are still left wanting.

The question itself is cumbersome. Where is the question coming from and what is being asked of Esposito? Is he expected to appeal to a prefabricated political roadmap *à la* Alain Badiou? Bosteels has argued that without such a gesture Esposito's notion of the "impolitical" leads into the apolitical terrain of the post-political. But is this a fair characterization? He is neither a vanguardist nor an avowed Maoist. Esposito fits within a clear trajectory of critical European political philosophers who prefer to search for new openings for rethinking radical politics by criticizing contemporary political formations rather than provide ready-made prescriptions. Other contemporary thinkers here include Giorgio Agamben, Jean-Luc Nancy, and Jacques Rancière. Each has been asked the same question. Some have provided clear answers, some elliptical ones, others have deconstructed the question itself.

Esposito has provided hints about his position, more so in recent years. We do not intend to speak for him; rather, we want to further situate his theory of community. In this article, we use his response in "Community, Immunity, Biopolitics" as a hermeneutical guide to examine his position on community in his trilogy (*Communitas*, *Immunitas*, and *Bíos*), in *Terms of the Political*, and in *Third Person*. We cannot possibly cover all the political ramifications of Esposito's work in a short paper. Instead, we want to engage in a thought exercise by drawing a comparison between his radical form of republicanism and the civic republican strains in the Anglo-American communitarian movement. We have chosen this route because we believe it helps to situate Esposito's theory in relation to mainstream politics. In other words, given Esposito's relative obscurity in mainstream political theory circles (particularly in North America), it is not only useful to situate his thought in relation to the dominant communitarian strands (and correlatively, to the liberalism to which it is inevitably connected). There is another reason as well: today in mainstream political discourse one cannot be taken seriously unless one acknowledges in advance that only some variety of liberal or communitarian thought is the basis for conversation. While Esposito brings in an element that (much like Nancy) is inevitably connected to the notion of communism, he does so in a way that establishes a filial connection to the mainstream liberal-communitarian tradition precisely by demonstrating that the latter "debate" shares the same premises, and that each fails in characteristic ways to go to the root (i.e., "radical" in the etymological sense) of the concept of association which they each presuppose. We thus also intend to demonstrate how his work radically reformulates some of the key tenets in mainstream political theory. We primarily focus on the relationship between community, freedom, and the proper. In the first section we examine how community is problematized in mainstream political theory. In the second section we begin with Esposito's critique of the mainstream framing of community, then we turn to his alternative solution of "affirmative freedom."

I the problem of community in mainstream political theory

A core presumption in the liberal approach is that the group is oppressive and individuals need to become autonomous agents to determine their own identity (see Esposito, *Third Person*; "The *Dispositif*"). Thus, liberalism, despite significant variations, takes as its fundamental value the autonomy of the individual. Berlin's now-famous distinction between two competing conceptions of liberty articulates how this autonomy has been conceptualized. The first of these, so-called negative liberty, holds that the necessary conditions of autonomy and freedom have been met when there is no external impingement on individual choice and action. The basic political thrust of negative liberty is to remove impediments to the sphere of individual human action. The second concept, positive liberty, holds that the mere absence of external impingement is not enough to assure conditions of autonomy and freedom. There are certain capacities, the possession of which is necessary (although perhaps not sufficient) for autonomy to be realized, such as equality of opportunity; this is something that mere formal equality under the law often cannot deliver. Accordingly, we might say that while negative liberty focuses on the formal or procedural conditions of freedom, trying to ensure that all are as free as far as possible from external constraints, positive liberty seeks to stipulate certain kinds of substantive freedoms the possession of which is needed for freedom and autonomy to prevail.

As is evident, and as Berlin argues in his preference for negative liberty, the realization of positive liberty for some will often require a diminishment of the liberty of others. Even though positive liberty might be desirable, it would seem that the only way to secure it would be to impinge on the sphere of freedom and autonomy enjoyed by others (for instance, requiring universal access to education means some would have to pay higher taxes, so that the loss of income would restrict their options, to some extent). Liberals have difficulty with such choices, since they would seem to require a clear criterion for making the necessary claim that some choices or values are more important than others, since this would clearly violate the necessary equation of freedom with autonomy of (individual) choice on which liberalism rests. For this reason liberals have often been charged by communitarians with espousing an inadequate and abstract concept of the subject and of the good (Taylor, "What's Wrong with Negative Liberty?").

Yet, far from being an admission of relativism or failure to reach consensus on the good (MacIntyre), liberalism can be seen as an epistemology containing a moral doctrine. In espousing individual autonomy and the equal value of the autonomy of each person, liberals are not simply giving in to relativism because they also believe (and here is their Enlightenment heritage), that in conditions of freedom, the truth will be attained. In other words, the corollary of individual autonomy is that this is the only way to attain an adequate conception of the good. The influence of bad or corrupt institutions that impose an obligation to obey established authority, individuals' dependence on a social order in which benefits are very unevenly distributed, and the undue power of received opinion backed by oppressive custom, are for liberals the great fetter to understanding, let alone attaining, the good. Nineteenth-century liberals, such as Mill, were doubtless far too optimistic in their assumption that the removal of oppressive custom would lead *automatically* to progress, truth, and the good, but were correct in believing that the latter could not be achieved if disagreement as to its nature or form was stifled in advance. Liberals of the classical era could also be faulted for their optimism with respect to the market and capitalism.

Under present conditions, this optimism with respect to the market also appears to be liberalism's Achilles heel. Indeed, if classical liberals could once imagine that the market would automatically harmonize the atomistic choices of free individuals in a manner productive of an optimal or even common good, the events of the last century and the beginning of this one have put paid to any such optimism. In our era, the formally free individual, with her "unconstrained" choice, becomes epiphenomenal to the combinatorial matrix of the global capitalist market; as several thinkers of the Frankfurt School asserted long ago, this puts the individual in an analogous position with respect to capital as that of early hominids confronted with the forces of nature. Today, enlightened reason – or as Sloterdijk would prefer, "cynical reason" – reverses into the hollowly superstitious optimism of the ideologically committed (those, in other words, with much more to lose than their chains). Under these conditions the liberal subject's dependency on a mechanistic framework of insulating and immunizing institutions becomes not only apparent but increasingly suggests that the subject of negative freedom, far from being the authorizing instance of such institutions, is in fact their idealized and ideological fiction.

This faith in the market is becoming liberalism's undoing in the contemporary era. That is, while it is easy to claim that liberalism as a whole is not equivalent to faith in the market, there is what Goethe would call an "elective affinity" between them. If the market is a kind of meta-coordinator of preferences, one allowing self-interested individuals to act freely and "spontaneously," it follows that for liberals the market is preferable to modes of regulation that require intervention by society (i.e., by the visible hand of the state) which would inevitably reduce the freedom of some (as noted above). If liberals don't have a clear set of criteria for limiting such freedom, the market provides an easy out: the meta-coordination supplied – or imposed – by the market allows liberals of various stripes to remain agnostic with respect to whose freedom should be reduced; if such reduction is merely an outcome of "spontaneous" coordination of choices, then *values associated with such choices need not be judged*. So the market provides liberals with an easy way out whenever they might otherwise be called upon to limit individual freedom. Further, given the elective affinity at stake here, it is quite tempting for liberals to view the market's coordination and limitation as something ultimately beneficial that works for the good and ensures overall benefit, if not progress.

Of course, this is not the end of the story. There clearly are situations in which no amount of market coordination yields satisfactory results, in which case it will be necessary to appeal to some set of higher-order principles that can be used to make decisions about the overall outcome of social interaction. Whether in the form of Rawls' principle of "justice as fairness" or some version of religious meta-ethics, liberalism is not in principle incompatible with a communitarian scheme that appeals to tradition, community standards, or even "our way of life" to decide conflicts between actors. But what would unite such communitarian versions of liberalism, or would seek to be compatible with liberalism, is an insistence on proceduralism. This is not to say, *pace* MacIntyre, that liberals are fundamentally committed to the right over the good, with the implication that liberals have no sense of the good, but rather that whatever communitarian values turn out to be appealed to in the case of conflicts between actors, such values must be formalizable in terms of rules. Otherwise, one is stuck with a kind of personal authority to intervene and decide such conflicts, an option that to the liberal mind is far too reminiscent of political theology and the divine right of kings for comfort.

Communitarianism represents a mainstream response to liberal defence of individualism. Communitarians contend that liberals *privilege* the individual to such an extent that communal bonds are lost. As we've noted, in a liberal democracy, as Isaiah Berlin famously claimed, a free person "should be left to do or be what he is able to do or be, without interference by other persons" ("Two Concepts" 121). Critiques of communitarianism, however, have argued that they have an "authoritarian tendency" (Esposito, *Communitas*), not unlike the one found in Rousseau's odd brand of republicanism, i.e., the "*forced to be free*" dictum (*Social Contract*). In response, communitarians claim that only their brand of liberty makes sense. Following Rousseau, they argue that the ideal of a pure form of negative liberty could only occur in a mythological state of nature ("Discourse"). Because we are social individuals, we can never be freed from external and internal obstacles. In fact, most communitarians attempt to salvage negative liberty by combining it with a social model of positive liberty. Charles Taylor's famous article "What's Wrong with Negative Liberty?" represents the general position held by communitarians on this issue: negative liberty (freedom from) is coextensive with positive liberty (freedom to do and become), i.e., establishing "collective control over the common life" ("What's Wrong with Negative Liberty?" 211). This mixed model of liberty is not unlike the dictum that "the free development of each is the condition for the free development of all" (Marx and Engels 15).

Communitarians also claim that the liberal emphasis on individual rights leads to a strong state and a weak civil society. The proceduralist model of liberalism gives rise to overly active juridical, legislative, and/or executive branches of the state. As an overarching, detached, and value-neutral arbitrator, the liberal state prevents the development of a commonly held value system. The state, in a sense, forces its negative freedom upon individuals and communities alike, which is a source of great tension in civil society. All that is left in civil society are rights-bearing private individuals who pursue their own interests.

Many communitarians thus prefer a weaker model of government and more autonomy for local democracy. In line with the civic republican tradition, they argue that community can only flourish when there is an active, self-governing, and relatively independent civil society. Following other republican thinkers, such as Rousseau (*Social Contract*) or more recently Arendt, communitarians elevate the role of political activity in the public sphere.

Some, such as Robert Putnam, cite Tocqueville's call for a strong and independent public in civil society, held together by voluntary associations; while others recognize that the public, meaning the people, has to be closely aligned with the state (Bellah et al.). Cultural communitarians, such as Charles Taylor, reduce community to cultural identity and then transfer this conflation into the political sphere as a national, ethno-cultural property belonging to a political community (Taylor, "Politics"; "Dynamics").

In generic terms, then, most communitarians appeal to positive liberty as an ideal for an active, or participatory, model of citizenship. In a civic republic, citizens are expected to *exercise* political control over their communal lives. Although participating in public life exposes individuals to their community, they stand to benefit more from their public acts than an immunized individual who lives a sheltered life in their own private spheres. Many trumpet the republican ideal of public life as the highest expression of individual life. Some, such as MacIntyre and Bellah et al., promote civic virtue, while others, such as Putnam, promote social capital.

II esposito's position

Rendered in his lexicon, the question about politics in Esposito's theory is posited as what an affirmative biopolitics would look like. What is the relationship between affirmative biopolitics and the common? In this special issue of *Angelaki* he argues that the "task at hand is to overturn in some way – indeed in every way – the balance of power between 'common' and 'immune,'" which consists of a delicate procedure separating the "immunitary protection of life from its destruction by means of the common; to conceptualize the function of immune systems in a different way, making them into relational filters between inside and outside instead of exclusionary barriers" ("Community, Immunity, Biopolitics" 87, 88). More particularly, we must contend with "two levels: by disabling the apparatuses of negative immunization, and by enabling new spaces of the common" (ibid. 88). Most of the work in Esposito's trilogy *Communitas*, *Immunitas*, and *Bíos*, as well as *Third Person*, focuses on the first level. He systematically deconstructs the ways in which the immunitarian dispositif operates within modern Western political theory. In the previous section of this paper, we briefly discussed the main contours in the liberal vs. communitarian debate. In this section we begin by outlining Esposito's criticism of this debate. Then we turn to Esposito's alternative proposition. What follows is just a rough outline of potential openings in his work for rethinking the common in an entirely counterintuitive fashion.

(a) esposito's critique

Esposito repeatedly claims that his theory of community differs from traditional accounts found in three representative branches of social and political theory: the organic model of *Gemeinschaft* in early twentieth-century German sociology, the Habermasian model of the communicative-ethical community, and the Anglo-American communitarian model ("neocommunitarian") (Esposito, *Communitas*; *Terms of the Political*; "Community, Immunity, Biopolitics"). Despite their differences, he contends, each conceives "of community in a substantialist, subjective sense" ("Community, Immunity, Biopolitics" 83). Because each is conceptually derived from the "figure of the proper," community is treated as a substance that belongs to the members. Membership is staked exclusively on each owner's claim over their commonality. What each *has* in common is "proper" only to those who belong to community ("Community, Immunity, Biopolitics"). This reading is not entirely new; it is likewise found in other texts written in the debate about community by Agamben, Blanchot, and Nancy (*Inoperative Community*; *Being Singular Plural*).

The new thinking of community is marked by a systematic attempt to reconceptualize the exclusionary and closed model. Each theorist argues that this closure is a by-product of the traditional tendency to reduce the common to the proper. Esposito's emphasis on the role of the immunitarian dispositif in the traditional configuration of community adds a new dimension to this discussion. Immunity is a terminological metaphor with several overlapping dimensions. Taken from the field of medicine, immunity indicates a function of the body, its capacity to insulate that body from a destructive external environment by identifying what is the body's *own* and eliminating or excluding that which is pathological to it. Already we have an organic dimension of the proper: what belongs to the body is that which contributes to its internal harmony, balance, and integrity, while what is improper is what breaks down or attacks such integrity. It is, to be sure, a very short step from this "proper" (literal) deployment of the concept to its metaphorical extension in politics. Here, the Greek notion of the city as the *polis*, whose citizen-members are constituted as a political body (i.e., a *politeuma*), displays the conceptual linkages between the city and its political members as a body, and, above all, a proper and "organic" body as natural as that of the biological body itself (see Aristotle, *Politics* Book I). And it is this body, both natural and political, that must be protected from what is external and improper. The immune paradigm is the mechanism through which the political body will be protected in a way analogous to the way in which the immune function protects the biological body. Meanwhile, according to Esposito's discussion in *Bíos*, what was already a mixed metaphor, that is, already biopolitical, has become explicitly so in the modern era. Building on the analysis provided by Foucault, Esposito argues that this primarily metaphorical sense of biopolitics has become literal in the application of new techniques of power to the life of populations. In this way it transcribes politics into a biological (and hence biopolitical) register which further expands the immunitarian aspects of modern civil order.

When this biopolitical and immunitarian defence is conflated with the proprietary model of community, members of a community who have exclusive rights to the common property establish collective mechanisms for protecting themselves. Internally, community shelters and protects insiders. Externally, it becomes a hostile and militant fortress. The ever-present threat of alien appropriation invigorates the immunizing apparatuses that create a mythological sense of unification. "[S]mall, micro-communities, opposed by definition to each other by ethnic, linguistic, and cultural identities" (Esposito, *Communitas* 54–55) become obsessed with containing their identities and whatever else "is considered to be properly their own" (Esposito, *Terms* 43). In our globalized era, these "besieged fortresses" (Bauman), whether in the Global North or South, have nearly "exasperated" the "autonimmunization" process to the point that common existence is no longer conceivable within the confines sketched out by traditional models of community (Esposito, *Terms*). This means, at least, that there can be no clear way to distinguish between inside and outside, us and them, because the inside is already marked by what it wants to exclude. This feature gives attempted reassertions of traditional modes of community membership a dangerously xenophobic character. The spiralling fear of "aliens in our midst" seems to necessitate ever stricter criteria of membership, reactivating the worst forms of racist, religious, and ethnic discrimination; we see this today when authorities are empowered to demand citizenship papers from anyone who looks like an "alien," as now happens routinely in parts of the United States and Europe.

Esposito argues, quite pointedly, that the "identity-making" communitarians ("Community, Immunity, Biopolitics") are at the forefront of the defence of the immunitarian model of community (*Communitas*). Their model of community is a by-product of the successive developments of "state sovereignty" and "individual rights," which are both at the forefront of the modern paradigm of immunization (*Terms* 128). After Hobbes, Locke, and Rousseau the "regime of the 'common'" was replaced by the regime "of the 'proper'" (ibid.). In biopolitical terms, each sphere becomes a form of property that must be immunized, often in contradictory ways, from external appropriation. The state and the individual become two privileged spheres, polarities, against which modern politics is determined. Modern political solutions, Esposito argues, are held by this "vise grip between public and private" ("Community, Immunity, Biopolitics" 89). Community withers into the background, is held back and repressed under the weight of the proprietary, sovereign, and hence privileged spheres of the public and the private. The communitarian attempt to foreground the public sphere thus becomes the mirror-image of traditional liberal commitment to the priority of the private sphere.

(b) esposito's solution

Esposito's notion of "affirmative freedom/liberty" represents a synthesis of his deconstruction of the proper and his critique of the immunitarian paradigm. How, he asks, can freedom be experienced in a community, when our prevailing models of liberty are bound up with a proprietary understanding of our world? This question appears in many of his writings, but for the purposes of brevity we will focus in this section on a short essay entitled "Freedom and Immunity" (in *Terms*). Esposito provides a whirlwind exploration of the problem of liberty in modern Western philosophy in this essay. Here, he states explicitly that if we are to solve the problem of thinking about community and freedom then we have to liberate "freedom from liberalism and community from communitarianism" (55). We begin by examining his statements about these two camps and then we end this section by examining his statements concerning affirmative freedom.

Mainstream liberal and communitarian models of liberty are derived from the metaphysical model of the subject. Both schools treat freedom as "a quality, a faculty, or a good" that a collective, an individual, or many subjects "must acquire" (50). The negative declension started appearing during the Middle Ages, when freedom began to be conceptualized as a "'particular right': an ensemble of 'privileges,' 'exemptions,' or 'immunity'" (52). This negative turn gave rise to a notion of liberty that exempts privileged subjects from common obligations and/or juridical responsibilities. Later political theorists, such as Hobbes, Locke, and Rousseau, inherited this model and cemented it as a cornerstone of modern political thought. In the negative model, subjects are expected to engage in a struggle to free themselves from any obstacle that could prevent them from becoming autonomous and proper subjects. Freedom is treated as if it is a *thing* that can be "appropriated" and thus "constituted as a subjective property" (50). Authentically free subjects must become fully-fledged proprietors of themselves, which means that one must become "'proper' and no longer 'common'" (ibid.). Translated into Esposito's lexicon, the private subject must be immunized from the common. In turn, the common becomes the anti-subjective realm that puts up communal barriers and responsibilities that negate the freedom of the subject. It comes to be viewed as the oppressive realm.

Rather than challenge the predicates supporting the negative model of liberty propagated by liberalism, communitarians make the sloppy mistake of merely translating this individualistic paradigm to the level of communities. "American neo-communitarianism," claims Esposito, mistakenly links "the idea of community to that of belonging, identity and ownership," meaning that community, as we have seen, is treated as a thing that people identify with "his/her own ethnic group, land or language" (48). Instead of creating an immunitarian enclosure around an individual, communitarians construct a perceived enclosure around the community. For them, community is concomitant with the proper. Communitarians treat the common as if it is "*one's own.*" Members of the community can only feel secure when their common property is immunized from external appropriation. These communities have built-in immunization mechanisms that are supposed to defend them from alterity. Thus, the negative inflection found in the liberal model of liberty is merely collectivized in this model, which is sometimes exaggerated to frightening degrees. In short, communitarians negate their own efforts to conceptualize a positive model of liberty that occurs within a group setting by reducing liberty to property and hence by reproducing the core logic of the immunitarian dispositif.

Esposito's effort to circumvent the closure of community found in the paradigmatic debate between liberals and communitarians begins with a turn away from the proper. Contrary to the communitarians, argues Esposito, community is "what is *not* one's own, or what is unable to be appropriated by someone" (49). Community can only be experienced as a "loss, removal, or expropriation" because it voids one's identity rather than fulfils it. Community is thus experienced as a radical alteration that disrupts the rigid boundaries that protect an individual's identity. Rather than appeal to an immunizing "enclosure" locking subjects inside themselves, either on an individualistic or collective basis, community is experienced in and through an opening and exposure that "turns individuals inside out, freeing them to their exteriority." His revision of positive liberty as "affirmative freedom/liberty" directly addresses this open model of community.

Esposito's deconstruction of the mainstream approaches to the relationship between community and liberty begins with a common gesture in his writing – he traces the origins of freedom back to the "semantics of community" (51). Prior to the negative turn, he notes, freedom was conceived as a relational concept with an "affirmative declension." With the Roman *libertas*, for example, freedom was configured "as the external perimeter that delimits what may be done from what should *not* be done" (52). Following the work of Jean-Luc Nancy, Esposito ties freedom to the singular plural model of existence. Freedom is not "something that one has," as if it were a thing one could appropriate, but it is merely "something that one is: what frees existence to the possibility to exist as such" (54). It is, following his work in *Communitas*, a practical experience in the "decision of existence." Freedom is nothing in particular. In fact, there really is "no freedom" per se, "only liberation"; "one cannot be" free because one "can only *become* free" (ibid.).

Without resorting to the evental tradition marking radical philosophical treatises on freedom in the twentieth century, Esposito attempts to "revitalize" the "affirmative power" of freedom by tracing it back to its "common root" where freedom is understood as the "locus of plurality, difference, and alterity" (55). This is the exact opposite of the modern sense of freedom as the "locus of identity, belonging, and appropriation." For Esposito, this means "freedom is the singular dimension of community" that "sweeps across infinite singularities that *are* plural" (ibid.). Translated into more practical terms, freedom can only be experienced in an open model of community that "resists immunization." This is a community that internalizes its exteriority while remaining open to difference. In this open and free community, individuals are exposed to alterity, pluralized, and thus prevented from appropriating differences.

Esposito's discussion of community represents a radicalization and revitalization of the idea of the common or the communal as it appears etymologically in Latin as *munus*. The *munus* as exposure to otherness and expropriating difference is not merely negative in the sense that the latter would represent a denial of something positive or good. Rather, nothingness as *munus*, as exposure to difference, is in itself something "positive" because it is how the relation that creates subject-positions becomes possible at all. Such negativity, although Esposito shies away from discussing it as anything but a lack – and this might well be a symptom of his insistence on de-substantializing communitarian belonging – in fact implies something of the positive freedom that communitarians seek but are inevitably led to reify in terms of the proper and the dialectic of belonging and disenfranchisement.

In any case, what is clear is that, as the *munus* implies, community takes the form of a collective debt that is owed to itself (and therefore anonymously), and it is here that Esposito's work resonates with the republican tradition. That the *munus*, as several contributors to this present issue will emphasize, is derived from the Roman Republic should be enough to establish Esposito's republican bent. But before one rushes to place at Esposito's door the defects of republicanism as it has been historically constituted – those precisely of which contemporary communitarianism seems most guilty – we must keep in mind that the *munus* is utilized by Esposito in order to enact a radical deconstruction of every notion of the proper.

Thus we believe that Esposito gestures towards and remains in proximity to a radical vision of republicanism. The radical in his theory lies both in the original sense of the term, found in his emphasis on the etymological origins of terms in his political lexicon, and in the contemporary sense of a politics beyond mainstream politics, which he usually defines as "impolitical." For Esposito, the *res publica* is neither "*la chose publique*," the "common wealth," nor a "common good" because the common is nothing but exposure to common being. The common is not a proprietary and/or moral good around which politics is circumscribed. Politics coordinated, mediated, determined, and grounded in things is closed and exclusionary. Politics, he contends, is ultimately grounded in nothingness. This traversal of the foundational elements of modern politics does not lead us to an abysmal form of absolute nihilism, which is a contradiction of terms; rather, it deconstructs the political terrain so that we are open to rethinking politics and ethics in an ontological manner. It is in this turning where we find his call for an affirmative model of biopolitics, which is as much political, as it is ethical, as it is communal.

It is with this point that we are led into the most difficult and problematic aspects of Esposito's work; these are aspects that it would be safe to say he is in the process of working out, and which form the centre of inquiry for several papers that follow in this special issue of *Angelaki*. Each paper, in its own fashion, puts a question to Esposito regarding the political significance, or practicality, of his theory of *communitas*; this is a task he too takes up in his contribution to this collection (see "Community, Immunity, Biopolitics"). What follows in this journal represents the most comprehensive commentary on his work in the English language to date. Although no article is without criticism, we hope that the reader will glean from this special issue a greater understanding of the important role that Esposito's political theory has within our contemporary era. We are thankful for all the strong contributions made.

Esposito raises a profound challenge to the near hegemonic domination that the proper has held over Western political thought beginning in the seventeenth century. Across his works one finds a consistent effort to expose the contours, dimensions, and significance of this crisis. Our intention in this short introduction is to further elaborate and situate his theory in relation to mainstream political discourse. Esposito provides a much-needed alternative perspective that helps us to think outside of the public/common vs. private property dichotomy that continues to shackle most political thought. Regardless of the political issue – immigration, national identity, economic imperialism, civil wars, international law, gender inequality – the discussion is framed by the proper. Competing claims are made on the basis of an individual's or group's proprietary rights over the issue. This now formulaic response logic has only exacerbated our pressing global problems. It reinforces the terms of the political that helped to define social inequalities, exploitation, and oppression, in the first place.

Alongside a handful of other contemporary political theorists that draw from different anti-proprietary strains in communist history, Esposito challenges us to rethink the hegemonic reign that the proper has over our contemporary capitalist sensibilities. Put differently, he re-raises the longstanding problem of thinking about political relations in terms of the market logic of ownership. His central concern traces back to the original enclosure movement that stole the commons from the people and replaced it with a hollow notion of public, state-owned property, up to our current intensive phase of privatization conducted under the auspices of neo-liberal globalization. If we are to combat this process and search for real, practical, political solutions, our politics must be conceived beyond the hegemonic reign that the proper has over our political contemporary political sensibilities. That is, if the proper appropriates and negates the common why do we continue to search for community through the purview of the proper? This question is of the first order for our era.

notes

We would like to take this opportunity to thank the editorial team at *Angelaki* for making this special issue possible. We are especially grateful to Charlie, Gerard, Harriet, and James, without whom this special issue would never have come to fruition. It has been a pleasure working with the editorial team at *Angelaki* and we look forward to possible future collaborations with everyone.

1 An alternative account can be found in Ignaas Devisch's paper that puts this in terms of rethinking the proper in an improper manner.

2 The place of politics in Esposito's theory has been raised from different perspectives by Bosteels, Campbell, and Neyrat. In this special issue of *Angelaki*, each of the authors has raised the question of politics from different angles, including Acosta, Bird, Deutscher, Devisch, Gratton, Hole, O'Byrne, Short, and Weir. The framing of the question is also the subject of the discussion that Esposito has with Jean-Luc Nancy in "Dialogue on the Philosophy to Come."

bibliography

Acosta, María del Rosario. "Hegel on *Communitas*: An Unexplored Relationship between Hegel and Esposito." *Angelaki* 18.3 (2013): 13–31. Print.

Agamben, Giorgio. *The Coming Community*. Minneapolis: U of Minnesota P, 1993. Print.

Arendt, Hannah. *The Human Condition*. Chicago: U of Chicago P, 1998. Print.

Aristotle. *Politics*. Trans. Ernest Barker. New York: Oxford UP, 1998. Print.

Bauman, Zygmunt. *Community: Seeking Security in an Insecure World*. Cambridge: Polity, 2001. Print.

Bellah, Robert N. et al. *Habits of the Heart: Individualism and Commitment in American Life*. Berkeley: U of California P, 2008. Print.

Berlin, Isaiah. "Two Concepts of Liberty." *Four Essays on Liberty*. Oxford: Oxford UP, 1969. 118–72. Print.

Bird, Greg. "Roberto Esposito's Deontological Communal Contract." *Angelaki* 18.3 (2013): 33–48. Print.

Blanchot, Maurice. *The Unavowable Community*. Trans. Pierre Joris. Barrytown: Station Hill, 1988. Print.

Bosteels, Bruno. "Politics, Infrapolitics, and the Impolitical: Notes on the Thought of Roberto Esposito and Alberto Moreiras." *New Centennial Review* 10.2 (2010): 205–38. Print.

Campbell, Timothy. "'Enough of a Self': Esposito's Impersonal Biopolitics." *Law, Culture and the Humanities* 8.1 (2012): 31–46. Print.

Deutscher, Penelope. "The Membrane and the Diaphragm: Derrida and Esposito on Immunity, Community, and Birth." *Angelaki* 18.3 (2013): 49–68. Print.

Devisch, Ignaas. "How (Not) to Properly Abandon the Improper?" *Angelaki* 18.3 (2013): 69–81. Print.

Esposito, Roberto. *Bíos: Biopolitics and Philosophy*. Trans. Timothy C. Campbell. Minneapolis: U of Minnesota P, 2008. Print.

Esposito, Roberto. *Communitas: The Origin and Destiny of Community*. Trans. Timothy C. Campbell. Stanford: Stanford UP, 2010. Print.

Esposito, Roberto. "Community, Immunity, Biopolitics." Trans. Zakiya Hanafi. *Angelaki* 18.3 (2013): 83–90. Print.

Esposito, Roberto. "The *Dispositif* of the Person." *Law, Culture and the Humanities* 8.1 (2012): 17–30. Print.

Esposito, Roberto. *Immunitas: The Protection and Negation of Life*. Trans. Zakiya Hanafi. Cambridge: Polity, 2011. Print.

Esposito, Roberto. *Terms of the Political: Community, Immunity, Biopolitics*. Trans. Rhiannon Noel Welch. New York: Fordham UP, 2013. Print.

Esposito, Roberto. *Third Person: Politics of Life and Philosophy of the Impersonal.* Trans. Zakiya Hanafi. Cambridge: Polity, 2012. Print.

Esposito, Roberto, and Jean-Luc Nancy. "Dialogue on the Philosophy to Come." Trans. Timothy Campbell. *Minnesota Review* 75 (2010): 71–88. Print.

Foucault, Michel. *The History of Sexuality, Volume I: An Introduction.* Trans. Roberty Hurley. London: Lane, 1979. Print.

Gratton, Peter. "Spinoza and the Bio-political Roots of Modernity." *Angelaki* 18.3 (2013): 91–102. Print.

Hole, Kristin. "The Ethics of Community: Nancy, Blanchot, Esposito." *Angelaki* 18.3 (2013): 103–18. Print.

MacIntyre, Alasdair. *After Virtue.* Notre Dame, IN: U of Notre Dame P, 1981. Print.

Marx, Karl, and Frederick Engels. "The Communist Manifesto." *The Economic and Philosophical Manuscripts of 1844.* Trans. Martin Milligan. Amherst, NY: Prometheus, 1998. 203–43. Print.

Nancy, Jean-Luc. *Being Singular Plural.* Trans. Anne E. O'Byrne and Robert D. Richardson. Stanford: Stanford UP, 2000. Print.

Nancy, Jean-Luc. *The Inoperative Community.* Trans. Peter Connor. Minneapolis: U of Minnesota P, 1991. Print.

Neyrat, Frédéric. "The Birth of Immunopolitics." Trans. Arne De Boever. *Parrhesia* 10 (2010): 31–38. Print.

O'Byrne, Anne. "*Communitas* and the Problem of Women." *Angelaki* 18.3 (2013): 125–38. Print.

Putnam, Robert. *Bowling Alone: The Collapse and Revival of American Community.* New York: Touchstone, 2000. Print.

Rawls, John. *A Theory of Justice.* Cambridge, MA: Belknap P of Harvard UP, 1971. Print.

Rousseau, Jean-Jacques. "Discourse on the Origin of Inequality among Men." *Basic Political Writings.* Trans. Donald A. Cress. Indianapolis: Hackett, 1987. 23–109. Print.

Rousseau, Jean-Jacques. *The Social Contract.* Trans. Maurice Cranston. New York: Penguin, 2007. Print.

Short, Jonathan. "On an Obligatory Nothing: Situating the Political in Post-Metaphysical Community." *Angelaki* 18.3 (2013): 139–54. Print.

Sloterdijk, Peter. *Critique of Cynical Reason.* Trans. Michael Eldred. Minneapolis: U of Minnesota P, 1988. Print.

Taylor, Charles. "The Dynamics of Democratic Exclusion." *Journal of Democracy* 9.4 (1998): 143–56. Print.

Taylor, Charles. "The Politics of Recognition." *Multiculturalism: Examining the Politics of Recognition.* Ed. A. Gutman. Princeton: Princeton UP, 1994. 25–73. Print.

Taylor, Charles. "What's Wrong with Negative Liberty?" *Philosophical Papers: Philosophy and the Human Sciences.* Vol. 2. Cambridge: Cambridge UP, 1985. 211–29. Print.

Tocqueville, Alexis. *Democracy in America.* Vols. I and 2. Trans. Henry Reeve. Toronto: Bantam, 2000. Print.

Weir, Lorna. "Roberto Esposito's Political Philosophy of the Gift." *Angelaki* 18.3 (2013): 155–67. Print.

maría del rosario acosta lópez

HEGEL ON COMMUNITAS

an unexplored relationship between hegel and esposito

foreword – an "impolitical" thinking: tracing the political in the shadows

According to Esposito, philosophy holds a paradoxical relation to the political: since it takes the political to be an *object* of thought, and therefore something apprehensible and resolvable through philosophical concepts, the political disappears every time it is translated into philosophical language. The inherent contradictions of the political, which remain unresolved if one does not want to uproot the political from what lies at its core, have traditionally been sublated by philosophy's tendency to "(dis)solve" or neutralize whatever it touches. Thus, "there is a direct relation between the *form* of political philosophy and its incapacity to think the political" (Esposito, *Confines* 20).[1] The political remains, in its most literal sense, *unthought*. Therefore, if there is still a task that remains to be taken up by political philosophy, it must be "only by means of its self-problematization"; only thus can political philosophy become "impolitical" and, hence, "think the political precisely where it shows itself to be irreducible to political philosophy" (Esposito, *Comunidad* 14).[2]

Even though contemporary thinking has repeatedly testified to this – Hannah Arendt and Simone Weil are obligatory references for Esposito (cf. Esposito, *Origen*) – philosophical thinking has yet to take it seriously enough. What remains, therefore, for political philosophy? It appears that every way out would end up trapped in the very same dialectic: to negate the political, or to approach it in its unrepresentability, would lead either to a negation of politics (the anti-political, writes Esposito, which coincides once again with the political in its own reversal (cf. Preface 102)) or to a negative political theology (cf. ibid. 104). It will not suffice either to point to philosophy's own limits or to try to operate from an "outside." If one truly wants to problematize political philosophy, one must do so from within:

> Every political concept shows an illuminated, immediately visible face, but also a dark zone that can only be drawn in contrast to light. One could say that modern political philosophy, blinded by this light, has completely lost sight of the zone of shadow that cuts across political concepts and does not coincide with their most manifest meaning. (Esposito, *Comunidad* 11)

There is an aporetic kernel inhabiting the shadow of every (modern) political concept, an unresolvable and multifaceted ensemble of meanings that ought to be unravelled in order to truly *resist*, and hence *interrupt* philosophy's tendency towards light, thus reopening a possible path from the political to thought. Therefore, thinking impolitically means to interrogate the tradition in a "transversal way," to "surprise it from behind" (ibid.). However, transcending political philosophy should not be understood here in contrast to an immanent move (Esposito, Preface 111). It means, rather, an "intensification and radicalization" of the political itself (ibid. 102) by means of an interruption of political philosophy *from within* that exposes it to "its *own* outside" (ibid. 111).[3]

Within this self-interpretive and deconstructive movement, the task of thinking community anew plays a decisive role. It demarcates its path and even constitutes its framework (cf. ibid.). In order to approach community impolitically, in order to seek a notion with the force required to interrupt its traditional established meanings, one has to deal "with an extremely subtle kind of thought" (Esposito, *Communitas* 15). Such thought would be one that, able to conserve the obliqueness and opaqueness inherent to its "subject," would also be able to trace the "broken paths" within tradition, the points where philosophy has moved to the verge of its own problematization, leading the way to impolitical thought.

This line of thought that runs "like a subterranean river through modern philosophy" (ibid.) is precisely the object of Esposito's 1998 work in *Communitas*. "Communitas" is hence the axis of suspension: a "point of departure, a hermeneutic support" (ibid. 3) that leads the way out of the traditional philosophical inversion of community into its opposite, both in its *immunitarian* version and in its *mythological* facet. Starting with Rousseau's response to Hobbes and following with Kant, Heidegger and Bataille, Esposito proposes the genealogy of a displacing and interruptive notion of being-in-common that, nonetheless, already lies at the heart of philosophical tradition. According to Esposito's reading, these thinkers, each in their own way, would have reintroduced "the question of community that modernity seemed to have completely closed off" (ibid. 15). In this way, each of them reopens community to a constitutive exposure against which modern political philosophy, starting with Hobbes, would have attempted to immunize it. Even if they all eventually succumb to the temptation of closing off what they had opened by falling back into a "mythological inclination" (cf. ibid.), these thinkers would have envisioned a way within (modern) political philosophy to escape from its apparently unavoidable dialectics.

I am convinced that Hegel is an obligatory stop along the way in this exploration. It is true that he has traditionally been read precisely as the paradigm of the opposite, the thinker of community as conciliation and totalization (cf. Esposito, *Communitas* 78–79, 104), as "re-appropriation" (cf. ibid. 100, 105, 115), and as the "thinker of the State," as Esposito frequently puts it (cf. *Terms* 19). And now that a certain trend of liberal political thought is "rediscovering" Hegel, he has been appropriated by a thinking of community that would still be trapped within metaphysics of the subject and which – according to Esposito – would still think our being-in-common within the paradigm of identity. This would be the case with philosophies of recognition and intersubjectivity, as well as with the liberal debate regarding communitarianism, to mention only two of the main examples that Esposito gives (cf. *Communitas* 2).

Nevertheless, it is precisely the fact that Hegel not only thought dialectically but also thought dialectics thoroughly – thus leading to its most definitive expression – that should capture our attention. Perhaps it is exactly at the point where dialectical thinking is driven to its limits that it may also implode, reaching its own limits, making them more explicit than ever, indicating the suspension that may be produced (with)in this very same movement. At the beginning of *The Inoperative Community*, and as an almost explicit wink at Hegel – via Bataille – Nancy writes:

> Excluded by the logic of the absolute-subject of metaphysics (Self, Will, Life, Spirit, etc.), community comes perforce to cut into this subject *by virtue of this same logic.* The logic of the absolute sets it *in relation*: [...] It undoes the absoluteness of the absolute. The relation (the community) is, if it is, nothing other than what undoes, in its very principle – and at its closure or on its limit – the autarchy of absolute immanence. (*Inoperative* 4)

If one follows such a suggestion, community is precisely the place where Hegel's thinking would be forced to interrupt itself. As Esposito points out, community is not just a place among others where the impolitical could be put into motion: *the interruptive gesture of the impolitical starts with and is constituted by a thinking of community.* Nancy's quote may be a good explanation of why this is the case, and as he develops it further throughout *The Inoperative Community* there is no one better than Bataille to point out this potency latent in Hegel's thought.

However, I would like to go further and affirm that this is not only a latent suggestion in Hegel that yet remains to be argued for by his readers. On the contrary, I will argue that Hegel is quite explicit about this possibility, precisely in connection with the notion of community, and thus continues the task Esposito has undertaken in *Communitas.* Following the very suggestive way in which Esposito deals with the question, and sharing his conviction that there is nothing more repressed and postponed than thinking community – and hence nothing more necessary and urgent (cf. Esposito, *Communitas* 1) – I want to show how Hegel's early thinking on community fits in a very appropriate way with this task. It is the task of looking for the points of suspension and inflection where *communitas* may obliquely and subtly reveal itself, even if faced with the inescapable risk of immediate negation. It is undeniable that whatever we read in Hegel today has been made possible through the language already brought to us by contemporary thought. It is definitely in tribute to this fact, and especially in reference to the way in which thinkers such as Esposito and Nancy have provided us with new language to think community anew, that I approach Hegel in this paper.

I will start by situating Esposito's reflections in *Communitas* within his broader, earlier approach to the problem of the "myth of community," as developed mainly in his 1993 book *Nove pensieri sulla politica.* This will give me a framework within which to explain in certain detail the most relevant aspects of his argument in *Communitas.* I will focus on how he describes the immunization paradigm of modern political philosophy, its (dialectical) relation to the "mythical reversal" of community, and the urgent task of interrupting these dialectics. The second part will start with a close reading of Hegel's criticisms of the violence of law in his early essay *The Spirit of Christianity and its Fate.* I will first show how the question of community appears in this essay in a context that is very similar to Esposito's analysis of the immunization paradigm and its mythical reversal. I will then proceed to outline how some of the main features of the "alternative" notion of community that emerges from Hegel's criticisms of law are closely linked – even in surprising and unexpected ways – to Esposito's reflections on *communitas.* I shall thus explain why Hegel occupies an important place in the deconstructive genealogy that gives rise to an alternative thinking of community as *communitas.*

communitas: from the sacrifice of community to the sacrificial myth of the common

the myth of community and the task of its interruption

Communitas begins by calling attention to a pervasive facet of community that, as Esposito argues, is not only expressed in twentieth-century totalitarianisms but even more so in contemporary political thinking (*Communitas* 1–2). This pervasive facet of community, which turns the political into the realm of the "common," appears as a reaction to the strong immunitarian force through which modern political thought defines itself. However, still trapped within the latter's conceptual framework, community is understood in this context in terms of totality, i.e., as the reversal and therefore the repetition of the absoluteness and totality of the individual. Therefore, instead of interrupting the immunitarian impulse, community conceived in this way extends this impulse to the whole of the social body. By describing this path from immunity to community in modern political philosophy, Esposito is also tracing the line *from* the *sacrifice of community* (immunity's impulse) to the *sacrificial myth of the common* (community's destiny). Furthermore, the genealogy of the book, from Rousseau to Bataille, delineates a thought of community capable of deconstructing this dialectic between the immunitarian paradigm and its "mythical reversal" (ibid. 16).

Esposito had already started to delineate this link between myth and the political in *Categories of the Impolitical*, which he then expanded and developed in *Nove pensieri sulla politica*. To interrupt the political, the impolitical must draw attention to the void of the political, i.e., to "the political but without its *mythico-operative fullness*" (Esposito, Preface 111; emphasis added). Like Nancy (*Inoperative*), Esposito argues that there is a straight line running from community understood as "Work"[4] – as a project, a destiny and an essence all realized *in common* – to its mythical face – which forces community to "identify with all of itself and with itself as all" in an "inequivocably totalitarian" gesture (Esposito, *Terms* 29). Contrary to our traditional reading, myth appears as the centre of political humanism, as part of its complete expression (cf. Esposito, *Confines* 97). The myth of the common – and thus, community as myth, its mythification and totalization – is itself the result of the idea of a *human* community, understood in terms of goal and Work. According to Nancy, myth is revealed in "the goal of achieving a community of beings producing in essence their own essence as their work, and furthermore producing precisely this essence as community" (*Inoperative* 2). Hence, even when myth is explicitly denied and questioned, it continues to operate – or operates even more than ever – with all of its force (cf. Esposito, *Confines* 100). Anticipating the (auto)-immunitarian structure that will become the centre of the analysis of modern political philosophy in *Communitas*, Esposito addresses this "logic of the myth" in relation to a dialectic where every negation is already trapped within what it intends to negate, thereby making it even stronger than before: "if there is anything the myth cannot tolerate, that would be its interruption, for myth is nothing else but the absence of interruption, or better said, the *uninterrupted itself*" (Esposito, *Confines* 109; emphasis added).

Myth and community are therefore linked in a dialectical movement that they both ground and nurture: "myth arises only from a community and for it: they engender one another, infinitely and immediately" (Nancy, *Inoperative* 50). Myth is thus always the myth of (a) community, which itself cannot but turn into the *myth of community*: "myth is always the myth of community, that is to say, it is always the myth of a communion – the unique voice of the many – capable of inventing and sharing the myth" (ibid. 51). Consequently, community becomes both the origin and the destiny of the common. Its mythological presence – and hence its totalitarian trace – operates every time community is understood as a (common) Work.

How then, Esposito asks, should this myth be interrupted? That is "what would be most demanded of us today: the interruption of the myth" (*Confines* 109). How could one, "even if just for an instant, close its mouth, always already opened to communication, to transparency, to the Work and its Completion" (ibid.)? If such a task is at all possible, it should not be understood as the end of myth but rather as a pause, as "a brief opening to the repose of the incomplete, to the shadow of the opaque, to the silence of the Work" (ibid.). The interruption of the myth is therefore its suspension, rather than a definitive interruption of its operation. "Stolen from the Work," this interruption would coincide, in its *inachèvement*, with the "form of a *désœuvrement*" (ibid. 110):

> The interrupted myth would therefore be the myth wounded in its pretense of a continuous complicity, of a reunification of voices, of a new community. This should not be understood in the sense of abolishing everything that is "common," thereby returning to a new individual Absolute, a new immanence of the Individual. Quite to the contrary, interruption here would mean reopening absolute immanence – of both the individual and the community – to the shared differences of a community of singularities. (Ibid.)

From Bataille to Nancy, rethinking community thus requires thinking the interruption and inoperativeness of community understood as totality, as work. This means not only to think it but also to *effect* it, if thinking here is not already the very gesture of such a *désœuvrement*. "Communitas" is therefore the gesture of this interruption: in keeping up with its potency both as radical ex-position and constitutive void, the thought of community attempts here to resist the temptation to fill the gap opened by the *munus*. By preserving the gap, the breach between community and thought starts to close, thereby inaugurating possibilities of thinking community anew.

To understand how *Communitas* undertakes this task, I offer a brief account of the steps Esposito takes to salvage a particular reading of *communitas* and the impulse to *immunitas* that such a *communitas* constitutively involves. By doing so, in the next section of this paper I want to present the main elements that will allow me to relate these two notions to Hegel's reading of community and law.

communitas, immunitas *and immunization*

In *Communitas* Esposito embarks on an expedition through the "genealogies" of this word as a "counter-history" of the traditional political concept of community. In his endeavour to displace the classic notion of community and substitute it with the original word "communitas," Esposito is taking account of the value and use of etymology to advance an idea that could offer resistance to the force of the inherited concept. Hence, this procedure should be understood as a search for a conceptual *topos* rather than an historical argument regarding its origin (cf. Manchev 22).[5] What does the word "communitas" entail that makes it possible both to *explain* and *resist* the tendency of community to completely disappear into its opposite (whether this be an absolutization either of the individual or of community itself)?

"Communitas" is first and foremost "community of law": there is a law that binds community together, an exigency that constitutes the locus of our very existence. However, this should be understood in a rather "Pauline" sense: "If we have always existed in the law, it is because we have always existed in 'guilt', one might say echoing Paul of Tarsus" (Esposito, *Terms* 15). That is because the *munus*, the duty and law that binds community together, is nothing but an obligation that, understood in its etymological sense, does not constitute us as subjects of rights, but even expropriates us from this subjectivity altogether. Following Simone Weil's criticism of rights,[6] Esposito writes: "to be subjected *to* an obligation means that we are subjects of nothing but our own expropriation: an expropriation of what is proper to us, beginning with the subjective essence" (*Immunitas* 23). The duty of the *munus* should be interpreted here in the line that goes from "*dovere*" (obligation) to "*donum*" (gift), and from "*donum*" to a very particular kind of gift: a gift that, still along the path of obligation, one cannot *not* give but for which, at the same time, one does not expect anything in return (cf. Esposito, *Communitas* 5). Gift as grace, thus, is understood as something one does not master but that one shares, in mutuality, only as a subtraction, as an "infinite lack," an always unsettled and "unpayable debt" (Esposito, *Terms* 15).

In *communitas*, therefore, subjects – already thrown into a radical impropriety – do not share "something" in common: the *munus* that binds them together is not a property, a content, even less a common essence or a project, but rather a void, "that distance, that extraneousness that constitutes them as missing from themselves;

'givers to' inasmuch as they themselves are 'given by' a circuit of mutual gift giving" (Esposito, *Communitas* 7). Even if, as Nancy rightfully clarifies, this should not be confused with a "revival of a certain Christian and humanistic emphasis on 'sharing', 'exchange' or 'others'" (Nancy, "Conloquium" 105), there is a path that Esposito wants to trace all the way back to the theological term *koinonia* and its Christian interpretation in the tradition of Paul of Tarsus (*Communitas* 10). *Koinonia*, the original term also used for the Eucharistic ceremony of "communion," distances itself from its immediate translation into *communio*, which is itself erroneously interpreted in a horizontal dimension of a "unity of spirits." Nothing could be further from its originary meaning, narrowly linked to that of *communitas*. If there is something "in common" in the Christian community it is already traversed by an infinite gift, an "otherness that withdraws us from our subjectivity [...] so as to pin it to a point that is 'void of subject,' from which we come and toward which we are called, just as long as we remain 'grateful' so as to respond to that first *munus*" (ibid.). To be "brothers in Christ" means to be bound by an infinite gift that interrupts any possible substantial bond, any possible appropriation: "'taking part in' means everything except 'to take'; on the contrary, it means losing something, to be weakened" (ibid. 11). There is an inflection of "participation" here, Esposito continues, that restores the echoes of the ancient expropriating *munus* in the Christian *koinonia* (cf. ibid.).[7]

This radical exposure experienced in/as *communitas* thus "interrupts the closing" of the subject within itself, and "turns it inside out" (ibid. 7). There is no-thing at the core of our being-in-common: the public-thing (*res publica*) shows itself to be in fact the no-thing at the ground of *communitas*, a groundless ground: "the breach, the trauma, the lacuna from which we originate" (ibid. 8). Nevertheless, to say "no-thing" is not to rehabilitate a negative ontology – which would itself fall back again into "the adoration of the nothing, of the absence, of the lack" which would end up again "covering the breach" (Neyrat 16). This "no-thing" "is not exactly nothing," as Nancy describes it in his introduction to the French translation of *Communitas*,

> [...] but something that is not a thing in the sense of "placed-present-somewhere." It is not in a place, since it is rather the place itself [...] the capacity to be found there with each other or among each other, the *with* or *among* being nothing other than the place itself [...] It indicates not a mystical nothingness but simply the *ex* that creates the exposition of existence. ("Conloquium" 106)

Hence, if no-thing binds us, there is also nothing that protects us from this exposure. The radicality of the *cum* means also that we are there, completely "in the open" ("Conloquium" 106), in a "violent loss of borders" (Esposito, *Communitas* 8). These borders are not only those of the individual: *communitas* does not enclose and protect the individual within the bounds of a collective realm. The borders are also those of community itself: being-in-common is also being "exposed to contact with the other, and thus to the possibility of a contagion from the other" (Esposito, *Comunidad* 16). That is why *communitas* is not only *at* permanent risk – at the risk of disappearance of, and immunization from, the *cum* – but it *is* also the risk itself.

This connects directly with the notion of *immunitas* and permits us to understand why *communitas* and *immunitas* are both at the groundless ground of our being-in-common. It is only because we seek to protect ourselves from our radical exposure to others that there is exposure and contact at all. Thus, Esposito does not seek to rescue *communitas* from the protective impulse of *immunitas*; both are equally constitutive and equally necessary for our existence. His criticism is rather of the tendency to endow *immunitas* with "the negative power of the negative" (*Communitas* 12), thereby turning it into "immunization." This is the move that becomes the "explicative key" and "paradigm" of the entire modern political reaction against *communitas*; the understanding of *immunitas* as the radical *opposite* of *communitas* situates both in a dialectical relationship that erases the original trace of their constitutive bond. Through this move, immunity is turned into the power (potential) of destruction not only of the effects of *communitas* "but even also of its presupposition" (ibid.). *Communitas* completely disappears in the face of a radical fear of death (Hobbes) that ends up sacrificing life itself to guarantee its own preservation: in the modern political paradigm individuals are thus "paradoxically sacrificed to their own survival" (ibid. 14). Hence, the *death of community* becomes a *community of death.*

This is not yet the end of the path described by modern political thinking; there is still one more step that perpetuates *communitas'* radical negation and turns it into its opposite. The immunization paradigm not only eradicates *communitas*; it also seeks to fill the void left by the negation of the *cum* with an even emptier bond. A radical negation of any relation (the individual as absolute) takes the place of the no-thing of the *munus*, thereby freeing each in advance from the debt and preventing any possible contact. The nihilistic character of this event nevertheless seems to bring its own dialectical move; a counter-reaction that attempts to fill the emptiness left by the absence of the bond. The claim for a "recovery" of community, therefore, takes place here as the claim for a *common* ground; a common essence and goal that instead of bringing *communitas* back into the equation in its "constitutively concave nature" displace it once again, this time by means of "its affirmative entification" (ibid. 15). The ab-solute of the individual is thereby transposed to the whole of community, and the "generality of the in-common" is hence reduced to "the specificity of a common subject" (ibid.).

Community's rebirth in the midst of modern political thinking is thus just a doubling of the immunization paradigm. Whenever this dialectic appears, we are at the core of the myth of community, of introducing community as a myth: as an origin one looks at with nostalgia, or a Work to be accomplished and achieved. Therefore, whenever this dialectical turn occurs, we are already in the midst of a drift imbued with a totalitarian impulse: "What was totalitarianism if not the phantasmatic temptation of filling in the fault, of closing the wound [...] of immunizing ourselves from the disease of melancholy by destroying its carrier germs?" (Esposito, *Terms* 30). This "mythical drift," Esposito remarks, accompanies any thought of community "like an originary risk [...] across the entire arc of its development" (*Communitas* 15).

Hence, we are now back where we started. The myth of community has managed to impede its interruption, precisely because any attempt to deny it is brought right back into its own dialectical flow. Is there, however, any thought that can linger long enough in the void of *munus* without falling back again into the temptation of filling it with content, of turning it into a myth once again? Community is both necessary and impossible for us, insists Esposito. Is it the same with the thought of community? Is there a thought that can bear both this necessity and this impossibility together? There seems to be a tradition within Modernity itself that opens the way to such a thought, one that would remain in the thought of *communitas* – at least long enough to envisage a way out of the sacrificial link between law and *munus*, between immunization and myth. Esposito has shown us that, among others, Rousseau and Kant had already started to *think this thought.* Would it be possible for us to find a trace of it in Hegel as well?

an interrupted community: hegel on immunization and *communitas*

In *The Spirit of Christianity and its Fate* (SC) Hegel delineates two steps that would be crucial in his later ethico-political reflections. On the one hand, he outlines a strong critique of modern ethico-political thinking and its correlate interpretations of community and law. On the other hand, this permits him to introduce an alternative notion of (ethico-political) community that he will continue to develop later on under the notion of *ethical life.* Hegel goes back to the history of religion to develop what J.M. Bernstein has recently interpreted as a "genealogy" of modern Western reason (cf. "Love" 397ff.), in its pervasive and destructive effects as well as in its latent, but yet to be fully understood, possibilities. Hence, as I intend to show, Hegel's reading of the step from a Judaic community-in-the-law to the early Christian life-in-common (with a very accentuated Pauline inflection)[8] follows the line from a life-protecting but also sacrificial sovereign law (that erases any possible communal bond) to a notion of being-in-common. The latter, in its resistance to the dialectics posed by the former, may approach that of Esposito's *communitas*.

I intend to show that one of Hegel's main concerns in the essay is the question of grasping the multiplicity and contingency involved in *life* (a central notion for his analysis of ethico-political community), without exercising and reproducing the very same violence and negative power that a universal and abstract standpoint (i.e., a "lawful sovereign" viewpoint) exercises over it. Thus, in this particular essay Hegel is not only concerned explicitly with the breach between (political) life and thought that Esposito reflects upon through his notion of an impolitical thinking but also very much aware that such a breach has been introduced and reproduced by thinking itself. Therefore, it is the task of philosophy to *interrupt* the violence of thought in order to open up a different path from philosophy to life and from community to law.

In what follows, I will only be able to delineate and indicate some of the passages where this is clearest in Hegel's text.[9] My interest is to show how, surprising as it may seem – especially considering more traditional interpretations of Hegel – Esposito's analysis in *Communitas* can shed a different light on the possibilities inherent in Hegel's approach to the subject. I will refer first to his reading of the "Spirit of Judaism," and reconstruct what he believes to be the inevitable consequences of a notion of sovereignty ultimately based on a "sacrificial" relation to law. Hegel's argument may be read as having the same structure as Esposito's analysis of the immunization paradigm, with its corresponding dialectics of totalization. I will then focus my attention on Hegel's analysis of early Christian community, as derived from Jesus' attempt to suspend the master–sovereign relationship to law and its totalization counter-act. Such suspension would be achieved not by negating them but rather by posing a certain interruption and unworking of their power. Thus, Hegel's essay would retrieve an alternative notion of community that may pose a resistance to the dialectics between *immunization* and the communitarian and totalitarian *myth*: a fragile and momentary notion of community, since it disappeared – according to Hegel – with Jesus' death.[10]

the violence of law: from the death of community to a community of death

I will start with a detailed reconstruction of Hegel's critique of the sacrificial paradigm of sovereignty and law, the first appearance of which, according to Hegel, is the "Spirit of Judaism."[11] Hegel starts his account of this Spirit's *destiny* in very similar terms to those that Esposito uses to describe the move of the modern political individual towards immunization. Everything starts with Abraham's "founding act." Guided by a need for absolute independence as well as an instinct of self-preservation, Abraham decides to radically cut himself off from every communal bond: "The first act which made Abraham the progenitor of a nation is a disseverance which *snaps the bonds of communal life and love* [and] the *entirety* of the relationships in which he had hitherto lived with men and nature" (SC 185, GC 245–46 [277]).[12] Every relationship is hence regarded as "hostile," as a possible threat to his independence.

Nonetheless, since this decision is guided by a need for self-preservation ("security," writes Hegel, "has been placed at the locus of existence"), it also comes with an individual self-subjection to an absolute external authority. However, this authority cannot itself be posed as a relationship. It has to be an absolute exteriority, an infinite objectivity, hence a "unity dominant over nature" that imposes itself in its absolute mastery (cf. SC 186, GC 246 [278]). Through this mastery Abraham attains his own, thereby guaranteeing his subsistence without the loss of independence: "His idea [the idea of God, of this infinite objectivity] subjugated the world to him, gave him as much of the world as he needed, and put him in security against the rest" (SC 187, GC 247 [279]).

This is the *fate* that Abraham has bequeathed to his people: "The relation of the Jews to one another was none other than the equal dependence of all on their invisible ruler" (SC 198, GC 255). The only possible "community" is one that comes out of a radical negation of every communal bond; the death of community, its *sacrifice* in exchange for the protection of the absolute external authority of the (divine) Law. Consequently, as in the case of Esposito, the move towards immunization is one that not only seeks to resist communal exposure but also destroys it altogether, leaving not even the trace of its presupposition (cf. Esposito, *Communitas* 12). Furthermore, for Hegel, as for Esposito, this will be Modernity's paradigmatic attitude towards the world: a reflective, nature-mastering reason, "originated" in this "absolute positing" of the Jewish God/Law, and clearly expressed in the turn represented by the political philosophy of Hobbes.

Thus, Hegel speaks in this context of a "universal enmity" (SC 191, GC 250 [283]), or – in a more explicit reference to Hobbes – of a universality of law that "presupposes equality, though the equality of enemies" (SC 216, GC 269 [328]). This is the main source of Hegel's criticism: the fact that this form of life is determined solely by an absolute (ultimately heteronomous) *legality*: "for they had committed [...] all love, spirit and *life* among men, to an alien object [...] what *held them together* was chains, laws given by the superior power" (SC 240, GC 290). Heteronomy, therefore, in exchange for "independence," legality and a universal (abstract) equality in exchange for *life* is, according to Hegel, what necessarily follows from that "infinite separation" posited at the ground of Jewish (ethical) existence.

Thus, by conceding all worth and power, love and life, to an absolute external Object, everything under its domain – and everything *is* necessarily under its domain – becomes worthless, "without intrinsic worth and empty, without life." "They are not even something dead" – Hegel continues – since "they are something only in so far as the infinite Object makes them something" (SC 191, GC 250 [283]). Being subject to this power thus means to sacrifice every form of life in exchange for a mere "physical dependence" – i.e., the sole kind of existence that "can be assured only *at the expense* [emphasis added] of all other existence" (SC 191, GC 250 [283]). Even at the expense of death, Hegel remarks: since that which decides on life, decides also on death. Existence under this Sovereign Law means being deprived even of the possibility of *death* itself. Or perhaps one should say that because one is deprived of the possibility of death, there is no real *life* under the law. Going back to Esposito and to his analysis of the *munus* at the groundless ground of being-in-common, one could also argue, this time with Hegel, that *communitas* has been sacrificed here since under the absolute mastery of Law one is deprived of the exposure that is existence itself. There is no existence anymore once life has been sacrificed to Law. Extirpating every *living bond*, every trace of a shared and henceforth exposed existence, Law demands the sacrifice of life itself in exchange for its protection. In Esposito's own words – as quoted above – individuals are thereby "paradoxically sacrificed to their own survival" (*Communitas* 14). Hence, Hegel also appears to be concerned with tracing the full circle that necessarily follows from the sacrifice of community to a community of sacrifice, or *from the death of community to a community of death*.

But Hegel is not only stressing how the paradigm of Law enters into the sacrificial circle. He *also* wants to show how this circle is inherent to the nature of Law itself, since this is the only way in which, due to its universality and abstractness, Law can *vest* itself with power, thereby legitimizing the violence that it henceforth exerts. This is Law's *hidden face* – or what remains in its *shadows*, as Esposito would put it: Law is absolutely impotent to claim its legitimacy if not by the exercise of this sacrificial violence that lies at the base of its own presupposition.[13] Hence, the law is sacrificial not only in its presupposition but also – and even more so – in its *implementation* – and in the attempt to safeguard its own capacity of implementation, its own power. Law has to sacrifice everything to ensure its own perpetuation; it has to negate *life* in order to preserve itself. There is thus a *doubling* of the sacrificial circle already implicit at the very core of the concept of Law.

In order to exert justice, Law is obliged to bind itself to what is most alien to it: "the law must be linked with life and vested with might" (SC 226, GC 279 [338]). This binding power is what the law lacks entirely in its mere concept. In the empty sovereignty exerted from its very own territory, Law is completely external and alien to the reality it must now intervene. Therefore, it is now forced to go through the experience of its own contradictory essence, since to actually *legislate* (to exercise its power) it has to abandon that which gives this power its legitimacy (its own universality): "there may be a contradiction between it as universal, as thought, and it as real, i.e., in a living being" (SC 226, GC 279 [338]). Such an experience, however, is not one in which Law arrives at its own destitution, thereby recognizing its own illegitimacy and the violence of its exercise. Inherent to its sovereignty and absolute mastery is also the stubbornness proper to its abstraction, to the empty unity of its idea. Law's own impotency, the weakness at the core of its own constitution, thus becomes the greatest of its dangers. There is hostility (*Feindseliges*) at the centre of Law's mastery over the world. Everything under its dominion has to be a confirmation of its power, but this also means that there is the potential of a threat in every single (concrete) being under its authority. Law cannot accept anything that contradicts its concept, and this leads ultimately to the worst kind of violence: that which is committed to disdain all reality unable to respond to (and therefore to *fulfil* and *confirm*) Law's own universality.

This is why Hegel describes the step from Law's presupposition to its empowerment as leading necessarily to a doubling of the sacrificial circle tending ultimately to the latter's final and most definitive (destructive) accomplishment. When the idea in the Law seeks to be actualized, whenever someone, in its name, feels "powerful enough to actualize its [Law's] idea of unity," this results in the merciless exercise of "the most revolting and harshest tyranny, and utterly extirpat[ing] all life; for it is only over death that unity hovers" (SC 289, GC 248 [279]). Hence, Hegel delineates here what he will later expand upon in the *Phenomenology of Spirit* (PS) under the "logic of terror": the disastrous consequences of an abstract empowered universality (one could also say a "fanatic" version of Modern reason) which, blinded by the vacuity of its conviction, oversteps any reality opposed to the idea it seeks to actualize. It is the terror of an "unbending cold universality" (PS §590, 525) that operates in ignorance of the one-sidedness of its concept: the emptiness and abstraction of its own perspective vs. the plurality and many-sidedness of the world in which it aims to immediately *actualize* its ideal. Nevertheless, precisely because of this emptiness, the universal Work of the idea never finds the correspondence it seeks in reality. That is why it can only present itself, time and again, as the "fury of disappearing" (PS §589, 525), as terror, as death's desolation, produced without the slightest remorse: "the coldest, emptiest death of all," – Hegel writes – "having no more meaning than chopping off a head of cabbage or swallowing a mouthful of water" (PS §590, 526).

Thus, "the universality of the concept is itself a *rending* of life" (SC 216, GC 269 [326–27]). Like the Gorgon's head, Law turns anything that dares to face it into stone, into something devoid of life:

> Outside the infinite unity [...] everything is matter (the Gorgon's head turned everything to stone), a stuff, loveless, with no rights, something accursed which, as soon as there is power enough, is treated as accursed and is then assigned to its proper

> place [death] if it attempts to stir. (SC 188, GC 248 [279])

Hence, Hegel is interested in taking the dialectics of immunization and its sacrificial circle one step further, to a second circle that appears also as an unavoidable result of the absolute foreignness and power of sovereign Law. The immunization paradigm is precisely what leads, correspondingly, to the operation of *totalization* (or better said, to totalization as *the* operation, the Work, of the Law). In Hegel's words, "so long as laws are supreme, so long as there is no escape from them, so long must the individual be sacrificed to the universal, i.e., be put to death" (SC 226, GC 278 [338]). The destiny of a "community-in-the-law" – a community that is posited in absolute bondage to Law – is not only to become a *community of death*, traversed by an economy and a *logic of sacrifice*, but also, ultimately, a community that sets death as its Work, that makes a work out of death. Under the infinite sovereign law, this community has one only goal: the *fulfilment* of the universality of its command even when everything singular must disappear in order to preserve this operation.

We are therefore, once again, in the midst of *myth*. In Nancy's words, "death, in such a community, is not the unmasterable excess of finitude" but rather "the infinite fulfilment" of the command of the Work (*Inoperative* 13). Hence, as is the case with the development of Esposito's argument in *Communitas*, Hegel has also succeeded in showing the path that leads from a community of sacrifice to sacrifice as the Work of community, from immunization against community to the gesture that turns community into myth. Incapable of bearing the void of the *munus* (to which existence-in-common is always already exposed), community-in-the-law fills the void of constitutive exposure with an even more negative emptiness. Such is the universality of Law's command, but this emptiness, as Hegel very well knows, is also the "fury" of totalization: the drive that seeks the content it both needs and constitutively lacks, therefore pursuing its own impossible fulfilment, destroying everything in its path.

Returning to Esposito's analysis, the question to be asked is: "how are we to fight the immunization paradigm without making it do death's work?" (Esposito, *Communitas* 19). How, indeed, are we to break down the dialectics between these two movements without falling back into either of them? The task at hand is to interrupt the myth of community without renouncing the thinking of a being-in-common, i.e., to remain in the difficult realm of accepting both the need of community and its impossibility. The question is whether Hegel himself might be able to say something about this possibility, now that we have seen how he has built an analogous context for such a question to take place at all. Such a community, if there is one, could not be one arising from the radical negation of sovereign law. The gesture, as seen above, has to be one of resistance from within, of interruption and suspension: a pause, Esposito says, that even if just for a moment opens up to the silence, the inoperativity that every Work carries within itself (cf. Esposito, *Confines* 109).

interrupting the violence of law: hegel's inoperative gesture

The question, therefore, is what sort of community may come out of the context that Hegel has delineated in his analysis of the Spirit of Judaism. Hegel is not interested in abolishing the Law: a "way out" of its violence has to come from within its realm, as resistance and not as dismissal. Hence, Hegel is much more interested in showing Law's weakness, its one-sidedness; he wants to bring to light not only the contradictions lying at its foundations but also the possibility of another relationship to the law, of another community that may remain at the same time within and beyond community-in-the-law. It would be a community, therefore, that may be able to open a breach at the heart of Law's own command, an impossible yet necessary community that will begin to die the moment it appears,[14] but will also open up a space in Hegel's thought for a very suggestive path towards a different being-in-common.

Finally, let me delineate how such a community may emerge, according to Hegel, from the early Christian notion of love and how this community-in-love will attempt to *interrupt* the dialectics between immunization and myth by *unworking* the link between the radical absence of community and the Work of its *totalization*. This gesture of interruption is signalled by what Hegel describes as Jesus' "sublime effort to *overcome* the whole of the Jewish fate" (SC 204–05, GC 261). Special attention will be paid henceforth to the two main aspects of this gesture: (i) what it means to be "together *in* love," in contrast to the kind of community united exclusively by (under) the (empty abstraction of the) Law; (ii) how one should read Jesus' attempt to "overcome" (*aufheben*) Law with his own "commandment of love," not through the negation of Law but rather through what Hegel, following Paul of Tarsus, describes as its "completion" (*pleroma*).

life-in-common or a community-in-love

To the enmity established by the Jewish spirit, Jesus will oppose the idea of a community "unified not in a universal, a concept (e.g., as believers), but *through* life and *through* love" (SC 278, GC 321 [393–94]): the love of *friendship* opposed to the "enmity" of Judaism, love as a *gift* (as the excess and overflow of life) opposed to the economy of sacrifice. It is still a community united in God, but no longer as a "collection of believers," i.e., no longer as individuals united only through their faith in an absolute external divinity, but rather as "a community *in* God" (SC 277, GC 321 [394]).

In order to fully grasp what this community *in* God entails, one has to understand the importance of the experience related to Christian love as well as the context in which Hegel's philosophy addresses religion. As J.M. Bernstein suggests, Hegel is interested in "decoding" an ethico-philosophical content implicit in both the Spirit of Judaism and Christianity. Furthermore, Hegel believes that Jesus himself was doing this, i.e., that he was also making manifest the implicit content of the concept of God in his teachings (cf. Bernstein, "Love" 395–96). Jesus is thus read by Hegel to be a kind of philosophical figure: one truly interested in overcoming the fate and exteriority of the Jewish Spirit, who would nonetheless have failed in transmitting this successfully to his own people. His community-in-love would again become a community of believers after his death, rendering this Spirit a "victim" of fate. However, what Hegel is interested in rescuing is Jesus' "reaction" to Jewish law in the form of a "commandment of love," and the very idea of community it brings about. If love is the place for community, the "common" is shattered here by an infinite gift that introduces God's radical alterity *within* community itself, as well as its in-appropriable excess at the core of our relationships to each other. This is what Esposito, also in relation to this "horizontal *communitas*" at the origins of Christianity (cf. "Flesh" 94), describes as the movement of "alteration and expropriation," lying at the "pre- (and thus also post-) Christian core that characterizes Christianity without coinciding with it" (ibid. 93).

Hence, whereas in the case of the Jewish Spirit God's radical alterity only shines forth in a community under the form of abstraction and death, in Christianity God is not only alive *in* community but immediately so, i.e., *as* community. Thus, Christianity is from the very first – and perhaps only at that moment – *a* community. After Jesus arrives, Hegel writes, "God is not isolated in anything but rather [is] *in* a living community" (GGC 311). God no longer designates an external concept, nor are relationships mediated by the presence of the divine as an absolute alien object. To live *in* God, to understand that the divine lives "in and among us," is to understand that such a community (*in* and *of* love) lives in the "*among*," in the passage leading from one to another and thus in the space that is opened in and through this exposure. Hence, love is "not the common character which a concept expresses" (SC 278, GC 321 [393]): it is neither a propriety nor a given (common) substance but a "living bond" (SC 278, GC 321 [393]). Due to an excess that no longer admits an empty "equality" or an exchange, love "overcomes [*aufhebt*]" the absolute oppositions that Law has established "as enmities" as well as the sort of "unifications of those subsisting oppositions" that Law produces under its (empty) concept of rights (SC 278, GC 321 [393]).[15]

Such an excess and expropriation that takes place in the experience of love is also what impedes understanding Hegel's community of love as an inversion of Jewish law's external alterity. For Hegel, Christian love – or better said, this early love characteristic of Christianity, even before it could be given the name – does not represent a turn towards an absolute immanence in which no real alterity takes place. If it did, the movement would be based on a simple replacement of one power and lordship over another. What takes place here is not an opposition between absolute transcendence and absolute immanence. The objective and positive alterity of law is assumed in a very particular way: *within* the living relationships (the many-sidedness of life) of Christian community. Since it begins precisely by questioning the very logic of the "proper," love – understood here, moreover, as an infinite gift – forecloses the "immanentist" and hence "totalitarian" logic of appropriation; it exposes and thus radically expropriates. Love takes place in the never-ending passage towards others, a passage always already opened to and by the alterity that traverses it.

The concept of love in Hegel's essay is the result of a development that had already started in "Die Liebe," another fragment written during the years in Frankfurt. It may be helpful to go back to this fragment to understand his argument about love as an interruption of the notion of "property." It goes without saying that this notion is at the centre of Esposito's discussion of community, and Bataille's notion of ex-propriation, which is also very close to Hegel's concept of love in his early essays. In the case of love, Hegel writes, "the giver does not make himself poorer; by giving to the other, he has at the same time and to the same extent enhanced his own treasure (compare Juliet in *Romeo and Juliet*: 'the more I give to thee, the more I have')" (Hegel, *Early* 307). The subject in love finds herself beyond herself; one could say it is already a *subject exposed*. Exposed by/to the other, it cannot but open further into this exposure. This radical exposure necessarily displaces, if it does not "annul" (by making superfluous, as Hegel will suggest) the very notion of the "proper." Judith Butler has also recently paid special attention to this early fragment in this very same direction: "the development of the phenomenon of love," she writes,

> involves a displacement of a purely subjective point of view – some dispossession of the self takes place in love. Internal to the singular and living feeling of love is an operation of life that exceeds and disorients the perspective of the individual. (11)

This notion of love, which in the fragment on "Die Liebe" is still related to an experience of losing oneself in the other ("consciousness of a separate self disappears" (Hegel, *Early* 307)), is replaced in *The Spirit of Christianity and its Fate* by love of friendship. It is still exposure and gift, but no longer an experience of losing oneself – nor one of gaining oneself in the process – but rather the kind of experience where community takes place by the very displacement of law's economic exchanges.[16]

This is not far from Esposito's notion of *communitas*, and even less from the Greek and early Christian notion of *koinonia* that he presents in his genealogical search (cf. Esposito, *Communitas* 10–11). Love is nothing but the place of exposure: it is neither a property nor a substance. What binds together here is openness itself, the "in" of the being-in-common, the infinite space (God) opened between each other. Hegel even introduces the notion of *gift* here, opposing it to the kinds of relationships established under the Law through the ideas of *debt* and *duty*: whereas a debt can always be settled, "purchasing thereby a favour by a sacrifice," love acts like a "warm-hearted gift, with no sense of duty or service" (SC 241, GC 291 [355]). Listening to the echo that resonates in the ideas of duty and debt, Hegel also decides to rescue the notion of *donum* here – of a gift that, within the realm of an (infinite) obligation towards others (evidently present in Jesus' teachings), does not let itself get caught in an economic exchange. Hence, as is also the case with Esposito's analysis and genealogy of the concept of duty, the "law of community" is rather this infinite and "unpayable debt" (Esposito, *Terms* 15); gift as pure grace opens the groundless ground of the common by introducing an "infinite lack." The law of community is hence nothing other than this lawless law that can be fulfilled only by subtraction, and can be shared only in mutual ex-propriation.[17]

love is the pleroma *of the law*

It is precisely following this trail that Hegel will explain the particular relation that such a community holds with the Law: "Only through love is the might of objectivity broken, for love upsets its whole sphere [...] Love alone has no limits" (SC 247, GC 296 [362]). Falling at the same time within and outside Law, Jesus' commandment of love breaks the Law from within, but necessarily exceeds it at the same time. Love interrupts because it *overflows* any possible symmetry, any logic of exchange, and any empty "equality" among the parties. Love has no limits, love alone does not have them, writes Hegel, and that is why it "upsets" the whole sphere from which the Law exerts its power, thus bringing along with it "another sphere of life" (SC 218, GC 271 [330]).

Jesus "took up a different attitude to those laws" (SC 209, GC 264 [319]). His commandment of love was intended to rise "above the whole sphere of justice or injustice" (SC 218, GC 271 [330]), both of which are still demarcated by law and therefore confirm its power. Love opens up an "aspect from which both it [the Law] and what conditions it can be seen to be subordinate to a higher sphere" (SC 228, GC 279 [340]). This sphere is thus not to be understood as the opposite of law: "The new life does not annihilate the content of the laws; it is, above all, its fulfilment [*Erfüllung*], the complement [*Ergänzung*] of what had existed until then in the form of something opposite, under the form of law" (GGC 311). Love does not interrupt the law by seeking to oppose it with another "unity," intended to replace the former and occupy its sovereign position; love interrupts the law because its presence fills the gap that gives law its legitimacy: love is, Hegel says, the *pleroma* of the law (SC 217, GC 270 [328]).

This *Erfüllung*, however, is again of a very particular nature, since it can no longer be the fulfilment that Law is always both seeking and failing to accomplish. The notion of *pleroma* that Hegel refers to here is taken from Paul's Letters in the New Testament (cf. Letter to the Romans 13.10; cf. also Luther 414ff.). It seems to indicate in the first place a relation of "supplement" that is able both to complement and to expose the deficiency of the Law. In his announcement of a "new right and a new light, a new region of life," Jesus does not disclose the fact "that laws must disappear"; instead, he tells his listeners "that they must be kept through a righteousness of a new kind [...] which is more complete because it supplements [*Ausfüllung*] the deficiency in the laws" (SC 214, GC 267 [325]). Therefore, *pleroma* here seems to designate the faculty that love (and only love) seems to enjoy, i.e., the faculty of complementing rather than merely "completing" the emptiness and "deficiency" of Law, which appears as an excess, an "expanded content," a "supplement" that breaks with Law's economic movement by introducing an "asymmetrical movement of a surplus" (cf. Hamacher 98). The fulfilment of Law, therefore, does not just simply fill the gap and emptiness left by the Law, but rather introduces a surplus into its heart that opens it from within.

By means of this very particular "operation," love *interrupts* the sovereignty of Law and Law "loses thereby its form as a law" (SC 214, GC 268 [327]). *Pleroma* indicates here a sort of de(con)structing relation to Law. Jesus' command "that ye love one another" is precisely the one command that, through its very form, *destroys* the meaning of command by taking it to its limits and leaving it groundless. Love cannot be commanded but it is, nonetheless, the one thing that Jesus commands. By setting it against the Jewish command "Thou shalt not kill," Jesus "not only does not act counter to this law, but makes it wholly superfluous [*überflüssig*]" (SC 215, GC 269 [326]). Thus, the poverty and deficiency of Law are exposed before the "richness" and "fullness" of life (cf. SC 215, GC 269 [326]). Such fullness, however, is nothing other than the "richness" of the lover, as mentioned above: a richness in dispossession, a fulfilment only in the sense of an exposure opened by the gift. Jesus thus "unworks" the law, making it "superfluous" by "displacing" [*verdrängen*] it with the richness of living relationships (SC 215, GC 269 [326]). This is the plurality of meanings that gives way here to the *Aufhebung* carried out by Jesus' gesture towards the Law, and hence to the idea of community-in-love. Could this be the beginning of a thinking of the sort of *désœuvrement*, influenced by Nancy, that inspired Esposito in the first place?

This relationship between the verb "*aufheben*" and the idea of an "inoperative" gesture is by no means arbitrary either, so let me just finish, in the spirit of Esposito's inquiry, with another etymological quest. In his reading of Paul's "Letter to the Romans," Giorgio Agamben notes how the Greek word Paul used to describe the relationship marked by the *pleroma* of the law is the verb *katargeo*, a "compound from *argeo*, which in turn is derived from the adjective *argos*, meaning 'inoperative, not-at-work (a-*ergos*), inactive.' The compound therefore comes to mean 'I make inoperative, I deactivate, I suspend the efficacy'" (Agamben 95). Furthermore, Agamben notes that Luther introduces the word "*aufheben*" in his German version of the Bible specifically to translate the Greek *katargeo* in the context of Paul's Letters (cf. Agamben 99).[18] Hegel does not use the word until *The Spirit of Christianity and its Fate*, where – as mentioned above – he is doing a close reading of Paul with the help of Luther's commentary. Therefore, the word "*aufheben*" appears in Hegel's work for the first time specifically in relation to this "inoperative" gesture that Paul attributes to Jesus' relation to Law. This "origin" of the use of the word in Hegel's work invites an in-depth revision of the way it has traditionally been interpreted in the context of his dialectics. This revision may well begin, as I have attempted to do here, with his notion of community which, along the line of thought leading from Paul of Tarsus to Hegel and from Hegel to Esposito, shows itself to be much more than a confirmation of its myth.

• • •

Much, of course, remains to be said about the possible relationship between Hegel and Esposito, and much remains to be questioned and problematized regarding it. If this study were meant to be a new chapter along the path that Esposito traces in *Communitas*, one would have to add a whole section on how Hegel's philosophy nonetheless inevitably becomes trapped in a metaphysics of the subject and a philosophical thinking that, as much as it opens up – if it does – another possibility for thinking community anew, it also helps, however, both to consolidate and reinforce the modern totalitarian notion of the common. How aware Hegel actually was of the risks his analysis so suggestively points out, how much he kept these questions alive in his later thought on community, and how much of this community-in-love would remain – for better and for worse – in his mature reflections on ethical life are questions that cannot be addressed here. Furthermore, how much of Esposito's own thinking is tacitly influenced by Hegel, both in his approach to the subject of community and in the way he develops it, is yet another question that I have had to leave aside for now.

My intention here has been much more modest, and my interest has been twofold: in the first place, I wanted to introduce Esposito's research on community within the broader problematic of the "impolitical" so as to call attention to the particular methodology he follows in *Communitas* as well as to show the relation between this work and some of his previous approaches to the subject. The question of community in this context appears closely linked to the task Esposito had already framed in previous works, i.e., the task of interrupting the *myth of the common* through its own de(con)struction. After clearly delineating Esposito's task in *Communitas* in these terms,

I was next interested in showing a close relationship between this task and methodology, and the way Hegel deals with the subject of community in *The Spirit of Christianity and its Fate*. Through this exercise, I hope to have shown how one can discover interesting hints of a more contemporary language of the impolitical in Hegel's work. Such an exercise is made possible in the light of Esposito's own language and the suggestive ways in which he approaches these questions in his work. Re-reading Hegel along the trail of these possible links may open up an interesting path for us in the present towards *another* way of thinking of community; one that may be able to interrupt the dialectics between immunization and the Work of death pointed out in Esposito's *Communitas*. Thus, I hope to have provided sufficient reasons to ascertain that Hegel may have played a crucial role along the path that leads from Rousseau to Kant in the genealogical exploration undertaken by Esposito in *Communitas*, and which prepares the way for Heidegger, Bataille … and Esposito himself.

notes

This work is the result of an encounter between two ongoing research projects: (1) the project "Narratives of Community: Politics and Violence" (2011–14), co-funded by Colciencias (convocatoria 521: Fondo Francisco José de Caldas) and the Universidad de los Andes; and (2) a book I am currently working on as a *Gastwissenschaftler* in Goethe Universität Frankfurt am Main, at the invitation of Professor Christoph Menke, funded by the Alexander von Humboldt Foundation. Some of the ideas that I discuss here were originally presented in the Eugenio Donato Seminar, Department of Comparative Literature at SUNY-Buffalo, in March 2012. I thank Rodolphe Gasché for inviting me to give that seminar, as well as all the participants for their insightful suggestions and questions. I also thank Félix Torres for his help in finding some of the bibliographical sources for this paper, and Sylvia Chaves and Ronald Mendoza for helping me with the final edition of the English version. Finally, I am also grateful to Greg Bird for the invitation to take part in this special issue of *Angelaki*: not only for his trust in my work but also for his support and endless patience.

1 Several books by Esposito have been translated into Spanish but have not yet been translated into English. In such cases I will provide my own translation of the text. This has also led me to present Esposito's thinking on community within the framework of some of his previous reflections, probably less known in the Anglo-American discussion of his thought: mainly *Categorie dell'impolitico* (1988) and *Nove pensieri sulla politica* (1993). I intend to show how *Communitas* (1998) participates in and continues with – but also deviates from and sheds a different light on – a project that Esposito had already delineated over ten years before its publication.

2 *Termini della politica. Communità, immunitià, biopolitica* (2008) has recently been translated into English (cf. Esposito, *Terms*). However, Esposito's original introduction is not included in the English version, so I must refer here again to the Spanish translation.

3 In a recent paper, Bosteels suggestively shows how this "impolitical" move in Esposito may again close the link between thinking and politics, giving as a result what Rancière has described as "metapolitics." According to Bosteels, what the impolitical "gains in terms of philosophical radicality," in its search for a deconstruction and a rendering inoperative of political philosophy, "it gives up in terms of political effectiveness" (Bosteels 237). I cannot deal with this criticism here, which may also have to do with a particular understanding of what the political and political action mean in the context of inoperative thinking. I will also show further on how Esposito does not want to naively go beyond politics into the impolitical, but is rather interested in the kind of resistance that can come, time and again, from a radical and continuous interruption of the "Work" of politics, without, however, pretending to "overcome" it or to point to a "truth" of the political that would lie behind politics.

4 Esposito capitalizes the word "Work" in this context to relate it explicitly to Lacoue-Labarthe and Nancy's analysis of the Romantic notion of Work (cf. Esposito, *Confines* 99). Cf. also their 1978 book on *L'Absolu Littéraire* (Paris: Seuil).

5 To deal with the question of whether Esposito's argument ultimately regresses to an "original sense

of community" (see also Manchev 22), one has first to understand what Esposito means by "constitutive" and "originary" when he refers to *communitas*. This also takes one back to a crucial aspect in the "impolitical" contemporary debate regarding community: the role that ontology plays in these reflections and how thinking community anew also means rethinking being. For a more detailed reflection on this exigency, cf. Acosta, "Tiny" 96–100.

6 Cf., for instance, Weil's wonderful book, originally published in 1949, *The Need for Roots: Prelude to a Declaration of Duties towards Mankind* (London: Routledge, 2002).

7 In a more recent article, written originally for a volume dedicated to Nancy, Esposito confirms again this close relationship between *communitas* and the Christian community at its origins. In the context of a discussion around Nancy's notion of the body and his deconstruction of Christianity, Esposito writes:

> The incarnation is the *munus* par excellence, the gift not only of life but also of the renunciation of the individual identity, at its origin the very idea of *communitas* as the dividing up of our singular and plural finitude. It is true that this horizontal communitas is destined to quickly roll over into an immunitary-type religio that makes the salvation of Christian society dependent on the preservation of its dogmatic-institutional form. (Esposito, "Flesh" 94)

Esposito proposes, therefore, to differentiate between two "movements" within Christian tradition: one that, in relation to incarnation ("Word made flesh"), "opens and exposes," and another that, in relation to "mankind's incorporation in the Church, and later in the State," "has instead the quality of unification and reappropriation" (ibid. 95). In a very similar gesture, as I will show below, Hegel will differentiate between the notion of community that results from Jesus' criticism of the Law, and the "destiny" of Christianity after Jesus' death (the "positivity" of the Church and its Institutions). In this sense, Hegel himself can also be read as the beginning of this "deconstruction of Christianity," or, as Esposito puts it in his reading of Nancy, of the reading of Christianity *as* deconstruction (cf. Esposito, "Flesh" 92).

8 In his approach to the subject and while writing the essay, Hegel was closely reading both Paul's Letters in the New Testament and Luther's extensive commentary on "Letter to the Romans." I must thank Jorge Aurelio Diaz for pointing the latter out to me (cf. Luther).

9 For a much more in-depth analysis of the subject, cf. Acosta, "La ley."

10 See n. 7 above.

11 For the time being I will leave aside what one could also call the "violence" intrinsic to Hegel's own criticisms, and the anti-Judaism present in his text. I would also like to stress the fact that, in his essay, Hegel genealogically traces the historical origins of modern rationality. Therefore, his assertions about Judaism are philosophically intended and not meant as value judgements (cf. also Ormiston 506).

12 I will quote directly from T.M. Knox's translation into English (SC), followed by the page number in Nohl's edition (GC) and also in Suhrkamp's revised edition, which will appear in brackets.

13 There is a connection here between Hegel's argument and Benjamin's diagnosis in his famous essay on *Zur Kritik der Gewalt*. There is also a very interesting link here between Benjamin's approach to the idea of sovereignty and Esposito's own analysis of the subject (cf. Esposito, *Confines*, especially the chapter on "Sovereignty").

14 As mentioned previously, this kind of community will not itself be exempted from fate. In Hegel's text, with and after Jesus' death the essence of community is again posited outside, in an external and future promise, instead of remaining within community as an infinite bond living in the *between*. The second part of Hegel's essay is therefore dedicated to exploring the *fate* of Christianity, but I will not elaborate on this side of the argument here.

15 Here I had to modify Knox's translation of this passage considerably: "Liebe [...] in der alle Entgegensetzungen, als solche Feindschaften, und auch die Vereinigungen der bestehenden Entgegensetzungen, -Rechte, aufgehoben sind."

16 I cannot elaborate further here on this particular notion of love of friendship, also always marked by the farewell of the other, like that which takes place at the last supper between Jesus and his friends. For a more detailed development of this topic, cf. Acosta, "La amistad."

17 For Hegel, whereas the relation to law, and hence to others under its command, can always be understood in terms of a debt that has to "be settled" – as is the case with punishment – the relation to others under the "commandment of love" introduces a radical interruption of this economy. Hegel speaks here henceforth of *forgiveness*, to show that there is nothing that guarantees the path towards others except for this infinite gift: an absolute thrownness for the other that can neither be commanded nor demanded, but only received. I cannot elaborate on these concepts here, but they constitute a whole different trail that can also shed light on the possibility of very interesting aspects of this unexplored relationship between Hegel and the thinkers of the impolitical community (cf. Acosta, "Variaciones," and idem, "La ley").

18 The word makes its appearance twenty-seven times in the New Testament, twenty-six of them in Paul's Letters.

bibliography

Acosta, María del Rosario. "La amistad como experiencia de reconocimiento: comentarios a una sugerencia de Hegel." *Amistad y alteridad*. Ed. Rodolfo Arango and Margarita Cepeda. Bogota: Universidad de los Andes, 2009. 97–108. Print.

Acosta, María del Rosario. "La ley como reproducción de la violencia: reflexiones tempranas de Hegel sobre comunidad, inoperancia e interrupción." *A la sombra de lo político: violencias institucionales y transformaciones de lo común*. Ed. M. Acosta and C. Manrique. Bogota: Sociedad Colombiana de Filosofía and Universidad de los Andes, 2013. 15–44. Print.

Acosta, María del Rosario. "A Tiny Displacement of the World: On Giorgio Agamben's Coming Community." *Epoché: A Journal for the History of Philosophy* 16.1 (2011): 93–112. Print.

Acosta, María del Rosario. "Variaciones sobre el perdón. Una sugerencia sobre política y transición a partir de Hegel." *Universitas Philosophica* 59.29 (2012): 33–50. Print.

Agamben, Giorgio. *The Time that Remains: A Commentary on the Letter to the Romans*. Trans. Patricia Dailey. Stanford: Stanford UP, 2005. Print.

Bernstein, J.M. "Confession and Forgiveness: Hegel's Poetics of Action." *Beyond Representation: Philosophy and Poetic Imagination*. Ed. R. Elridge. Cambridge: Cambridge UP, 1996. 34–65. Print.

Bernstein, J.M. "Love and Law: Hegel's Critique of Morality." *Social Research* 70.2 (2003): 393–432. Print.

Bosteels, Bruno. "Politics, Infrapolitics, and the Impolitical. Notes on the Thought of Roberto Esposito and Alberto Moreiras." *New Centennial Review* 10.2 (2010): 205–38. Print.

Butler, Judith. "To Sense What is Living in the Other: Hegel's Early Love." *100 Notes – 100 Thoughts*. No. 66. Kassel: dOCUMENTA (13) and Ostfildern: Hatje Cantz, 2012. Print.

Esposito, Roberto. *Categorías de lo impolítico*. Buenos Aires: Katz, 2006. Print. [Original Italian title: *Categorie dell'impolitico*. Bologna: Il Mulino, 1988.]

Esposito, Roberto. *Communitas: The Origin and Destiny of Community*. Trans. Timothy Campbell. Stanford: Stanford UP, 2010. Print. [Original Italian title: *Communitas. Origine e destino della comunità*. Turin: Einaudi, 1998.]

Esposito, Roberto. *Comunidad, inmunidad y biopolítica*. Madrid: Herder, 2009. Print. [Original Italian title: *Termini della politica. Communità, immunitià, biopolitica*. Milan: Mimesis, 2008.]

Esposito, Roberto. *Confines de lo político. Nueve pensamientos sobre política*. Trans. Pedro Luis Ladrón de Guevara. Madrid: Trotta, 1996. Print. [Original Italian title: *Nove pensieri sulla politica*. Bologna: Il Mulino, 1993.]

Esposito, Roberto. "Flesh and Body in the Deconstruction of Christianity." *Minnesota Review* 75.1 (2010): 89–99. Print. [Original French title: "Chair et corps dans la déconstruction du Christianisme." *Sens en tout sens. Autour des travaux de Jean-Luc Nancy*. Ed. F. Guibal and J.C. Martin. Paris: Galilée, 2004. 153–64.]

Esposito, Roberto. *Immunitas: The Protection and Negation of Life*. Cambridge: Polity, 2011. Print. [Original Italian title: *Immunitas. Protezione e negazione della vita*. Torino: Einaudi, 2002.]

Esposito, Roberto. *El origen de la política ¿Hannah Arendt o Simone Weil?* Madrid: Paidós, 1999. Print. [Original Italian title: *L'origine della politica. Hannah Arendt o Simone Weil?* Rome: Donzelli, 1996.]

Esposito, Roberto. Preface to *Categories of the Impolitical*. Trans. Connal Parsley. *Diacritics* 39.2 (2009): 99–115. Print.

Esposito, Roberto. *Terms of the Political: Community, Immunity, Biopolitics.* Trans. Rhiannon Noel Welch. New York: Fordham UP, 2013. Print.

Galli, Carlo. Review of Esposito's *Nove pensieri sulla politica. Filosofia politica* 1 (1994): 154–56. Print.

George, Theodore. "Forgiveness, Freedom, and Human Finitude in Hegel's *The Spirit of Christianity and its Fate.*" *International Philosophical Quarterly* 51.1 (2011): 39–53. Print.

Hamacher, Werner. *Pleroma: Reading in Hegel.* Stanford: Stanford UP, 1998. Print.

Hegel, G.W.F. *Early Theological Writings.* Trans. T.M. Knox. New York: Harper, 1961. Print.

Hegel, G.W.F. "Geist des Christentums" [GC]. *Hegel theologische Jugendschriften.* Ed. Herman Nohl. Tübingen: Mohr, 1907. Print.

Hegel, G.W.F. "Der Geist des Christentums und sein Schicksal" (page numbers appear in brackets). *Werke in 20 Bänden.* Frankfurt am Main: Suhrkamp, 1970. 274–417. Print.

Hegel, G.W.F. "Grundkonzept zum Geist des Christentums" [GGC]; and "Geist des Christentums" (page numbers appear in brackets). *Werke in 20 Bänden.* Frankfurt am Main: Suhrkamp, 1970. 297–316. Print.

Hegel, G.W.F. *Phenomenology of Spirit* [PS]. Trans. Terry Pinkard. 2008. Web. <http://web.mac.com/titpaul/Site/About_Me.html>.

Hegel, G.W.F. *The Spirit of Christianity and its Fate* [SC]. *Early Theological Writings.* Trans. T. M. Knox. New York: Harper, 1961. 182–301. Print.

Luther, Martin. *Comentarios de Martín Lutero. Carta del apóstol Pablo a los Romanos.* Trans. Erich Sexauer. Barcelona: CLIE, 1998. Print.

Manchev, Boyan. "Puissances de la communauté." *Collège International de Philosophie* 59 (2008): 17–22. Print.

Nancy, Jean-Luc. "Conloquium" (preface to the French Translation of Roberto Esposito's *Communitas*). Trans. Janell Watson. *Minnesota Review* 75 (fall 2010): 101–07. Print. [Original title in French: "Conloquium." *Communitas. Origine et destin de la communauté.* By Roberto Esposito. Paris: PUF, 2000. 1–10.]

Nancy, Jean-Luc. *The Inoperative Community.* Trans. Peter Connor. Minnesota: U of Minnesota P, 1991. Print.

Neyrat, Fréderic. "Immunité globale." *Collège International de Philosophie* 59 (2008): 13–16. Print.

Ormiston, Alice. "*The Spirit of Christianity and its Fate*: Towards a Reconsideration of the Role of Love in Hegel." *Canadian Journal of Political Science* 35.3 (2002): 499–525. Print.

greg bird

ROBERTO ESPOSITO'S DEONTOLOGICAL COMMUNAL CONTRACT

> The distinguishing feature of Communism is not the abolition of property generally, but the abolition of bourgeois property. But modern bourgeois private property is the final and most complete expression of the system of producing and appropriating products that is based on class antagonisms, on the exploitation of the many by the few. In this sense, the theory of Communists may be summed up in the single sentence: Abolition of private property.
>
> *Marx and Engels, "The Communist Manifesto" 223*

> [S]ystematic communism, the deliberate negation of property, is conceived under the direct influence of the prejudice of property, and it is property that is to be found at the root of all communistic theories.
>
> The members of a community, it is true, have no private property, but the community is the proprietor not only of goods but of persons and wills. It is because of this principle of absolute property, that labour, which should be a condition imposed on man only by nature, becomes a human commandment [...]
>
> *Proudhon,* What is Property? *196*

Roberto Esposito's deconstruction of the "semantics of the *proprium*" (*Communitas* 2) directly addresses these competing perspectives on community. Besides loose etymological analyses by anarchists like Pierre-Joseph Proudhon and Max Stirner, both criticized by Karl Marx and Frederick Engels (*German Ideology* 245–48), this debate never carried beyond the core distinction between private (or individual) vs. public (or common) property. The modern rendition of the semantics of the proper began when the "West" set out to colonize the "rest" of the world. From Pope Alexander VI's Papal Bull of 1493, to the enclosure movement in England, to the social contract tradition epitomized by John Locke's theory of property (*Second Treatise*), to contemporary rights discourse, to trade-related aspects of intellectual property rights, the West has been engaged in a slow process of converting everything in its wake into private property.[1]

This process extends beyond the "primitive accumulation" of private property. Most contemporary conservatives, liberals, and communists are committed to the hegemonic discourse that I call "the political economy of the proper" in this paper. Proudhon pointed to this discourse when he coined the term "property prejudice," but it wasn't until Martin Heidegger set out to wrest ontological philosophy away from the semantics of the proper, property, propriety, even authenticity (all covered in his writings on *das Eigen*) and Derrida carried this project forward in his multi-volume analysis of the proper (best exemplified in *Specters of Marx*), that the hegemony of this political, and for some theological, economy began to be seriously challenged. The texts in the recent Continental debate on community written by Esposito (*Communitas*), Giorgio Agamben, Maurice Blanchot, and Jean-Luc Nancy (*Inoperative Community*; "La Comparution"; "Of Being") have brought this problematic to the forefront of social theory.[2]

Esposito's *Communitas* adds a unique dimension to the contemporary debate concerning community. His work synthesizes the communitarian critique of the social contract tradition, Heideggerian existentialism, and the communist question. Politically, he prioritizes communal duties and obligations over private civil rights, interests, and property. In his rendition of communism, ethics must supersede economics. Existentially, his community unfolds in an ontology where *being* takes precedence over *having*. Rather than replace the contractual model altogether, Esposito conceives of a heterodox model of contractual exchange that I call a "deontological contract." It is deontological in two senses. First, it is *deon*tological because the *munus* of *communitas* carries the force of an ethical duty. Second, it is de*onto*logical (with the "de-" representing an intensive prefix) because it occurs in and through the event of existence. For his contractual exchange to be effective, it can only occur as a disruption of the political economy of the proper. This paper consists of an examination of his efforts to bring these disparate traditions together under this single problematic. The order of this article reflects the movement in *Communitas* from the early chapters on contractualism and duty ethics to his later chapters on existentialism, ontology, and nihilism.

part i: the dialectic of alienation and appropriation

Esposito's theory of community is drawn from the tradition of emancipatory politics. He asks: how can we conceive of an emancipatory community that is no longer proper? To do so, community would have to be understood as improper (with the "im-" standing for both "inside" and "against" the proper).[3] In *Communitas* this work is conducted across multiple fronts. For introductory purposes, I begin by examining Esposito's deconstruction of the dialectic of alienation and appropriation. This subtle thread in *Communitas* is used by Esposito to bring existentialism and the social contract tradition together.

In proprietary renditions of emancipation, a community is constituted through the act of negating, overcoming, and appropriating previously alienating conditions. A proper community can only be constituted when subjects transform their collective state of being not-property into property. This simple formula of negating their negation, their lack, is repeated in most theories of collective emancipation, including anti-imperialist, national liberation, feminist, queer, anti-racist, and communist politics. For each, belonging is calibrated according to the political economy of the proper. This could take the literal form of economic property, or in the abstracted forms of personal identity (rights discourse), ideas (self-determination, autonomy, or patents), bodies (abortion, euthanasia, or aesthetic freedom), even collective identities (strategic essentialism, national identities, or cultural patents).[4]

In the literature on community, this dialectic has been attacked from many angles. It is based in the metaphysics of the subject, it reduces difference to sameness (or identity), it is trapped in the onto-theological search for an original foundation, it creates an inclusion/exclusion framework, etc. Each theorist comes to the same conclusion: this dialectic prevents us from thinking community because it produces closed relationships, which are ultimately not relationships.

The dialectic of alienation and appropriation also leads us to treat community as if it were a thing. Traditionally community is treated as "the thing itself that is opposed to its own destruction" (*Communitas* 136). As a thing it is supposed to *immunize* us from the "explosion (or implosion) of the nothing," i.e., from nihilism (ibid.). Esposito argues that this immunization imperative actually results in a nihilistic model of community. His argument, which is largely drawn from Nancy and others in the tradition of existentialism, is astute. When community is hypostatized and subjected to the dialectic of alienation and appropriation, we are expected to commit a two-fold act of appropriation to qualify as proper members of a community. By characterizing community as a thing, and not just a simple nomination for relationships, each is expected to appropriate not only property but also the community itself. This second appropriation constitutes the right, even privilege, to claim partial ownership over the community. Without this second appropriative act each would be merely engaged in an individualistic act of appropriation, which is the commonplace characterization of nihilism.

Nowhere is this collective nihilism more apparent than in the tautology of participation and sharing that factors into most revolutionary formulas. Each revolutionary member must simultaneously participate *in* the taking and take part *of* the participative activity. The former requirement is easier to comprehend: each must be a participant in the taking of expropriated materials, alienated identity, the state, etc. Each takes their share of, appropriates, what had previously been taken away from them. As an end in itself, this activity is not conducive to community because it is conducted primarily for private purposes. For example, during riots, such as those in England and Vancouver in 2011, participants engaged in fleeting appropriative acts that were not translated into a properly communal action. They sought neither to transform their relationships with each other into communal relations nor to transform their broader circumstances of shared privation into communal conditions. They merely inverted the political economy and dramatized it on a collective scale. The resulting mayhem simply reconfigured the already divisive conditions that are a by-product of the collective nihilism that marks our times.

In the standard narrative, nihilism is contained when taking occurs on a communal scale. Each participant must appropriate the collective activity and make it proper to herself, i.e., to appropriate her relationships, to take ownership over them, and to become a co-proprietor of her relationships with others. Everyone must *have a share* in the participative activity itself. Sharing is the modality, the basis, and the bond of communal relationships. Without this shared element – sharing out, sharing in, and ultimately sharing with – appropriation remains uncommon. Each engages in a negative act of appropriation that is private, individualistic, and anti-social, or in his terms immunitarian. Esposito claims this merely amounts to a "division without sharing" (28).

The question that Esposito – and Nancy before him – force us to address is: does this two-fold appropriation absorb and nullify the division that defines our commonality? That is, does it lead to what we could call a "sharing without division"? But this raises another question: how can each participant appropriate the community without annihilating it, or, conversely, how can each participant be appropriated by community without being completely absorbed, and thus annihilated, by it? This aporia is found in all zero-sum formulations of this problematic that are promoted by those who appeal to the metaphysics of the subject. Sartre's exhaustive, yet failed, efforts in the *Critique of Dialectical Reason* testify to this aporia. To appropriate relationships, to make them each and everyone's own, to make them proper to one's person and every other person involved, to qualify them as "ours" and to enter into a "we," is to negate the very exteriority that is required to be in – rather than have – a relationship with an other in the first place. Appropriating relationships, in short, renders them non-relational, which is why Esposito calls Sartre's work a "great communitarian failure" (*Communitas* 133). The political economy of the proper only gives us two alternatives to this dilemma: immunized, private individuals found in the social contract tradition (division without sharing) or a hypostatized community that absorbs its subjects (sharing without division).

Étienne Balibar has praised Esposito and Nancy for their rigorous deconstruction of the role of property in the construction of political borders in an essay called "Citizenship without Community" in *We, the People of Europe?*, but elsewhere he has argued that thinkers who seek to deconstruct the proper, notably Derrida, effectively promote a model of politics that radicalizes alienation and thus leads to an "*abyssal alienation*" ("'Possessive Individualism'" 315). Drawing from the same logic, critics of Esposito and Nancy have charged both thinkers for scribing a theory of community that is politically impractical.[5] These criticisms miss the core argument made by both Esposito and Nancy: each argues that alienation is intimately bound up, meaning inseparable from and co-original, with the proper. Appropriation, private or common, is thus not an adequate solution to alienation.

Our era is marked by the excessiveness of the property prejudice. The political has been usurped by the economic. We usually refer to this as the predominance of economic interests in the political sphere, but the problem is more widespread because today the political is now defined by the economic. Political rights, for example, are conceptualized in terms of property. Substantial political rights, active citizenship, is distributed on the basis of owning large sums of economic property; yet the allocation of formal political rights is likewise predicated in terms of political property: ownership over one's own, collective or individual, person (Esposito, *Third Person*). Although neither thinker develops a practical alternative "model," which would be contrary to their core values, each has provided us with many practical insights into what happens to radical models of politics when they are framed by the political economy of the proper. For the remainder of this paper, I focus on Esposito's efforts to open up new trajectories for rethinking community as a disruption of the political economy of the proper.

part ii: *deon*tology and the *munus*

> Modern individuals truly become that, the perfectly individual, the "absolute" individual, bordered in such a way that they are isolated and protected, but only if they are freed in advance from the "debt" that binds them one to the other; if they are released from, exonerated, or relieved of that contact, which threatens their identity, exposing them to possible conflict with their neighbor, exposing them to the contagion of the relation with others. (Esposito, *Communitas* 13)

In the introduction to *Communitas*, "Nothing in Common," Esposito meticulously examines the displacement of community in political philosophy. Community, he argues, "isn't translatable into a political-philosophical lexicon except by completely distorting (or indeed perverting) it" (1). Modern political philosophers treat community as if it is a "property," a "thing," a "subject," a "substance," and most concernedly a "'wider subjectivity'" that "swells the self in the hypertrophic figure of 'the unity of unities'" (1–2). What if, he asks, it is not any of these things because that which *draws us together* is nothing?

Esposito argues that we need to wrest community from the social contract model because it conflates the political with the economic such that politics become commercialized. What if, he asks, the communal contract was conceived not in these terms, but as a deontological contract? What would remain were community not reduced to the economic, the political, or political economy in general? Esposito turns to the forgotten etymology of *communitas* for answers. *Com-munus* is a mere contract that draws us together in an improper form of exchange. It is neither a thing itself nor a contract that we enter into in exchange for something in return. All that is exchanged in this contract is *ourselves*. When we are drawn together in the contract of the *munus* we are expropriated so extensively that we are rendered incapable of appropriating the contract, others, even ourselves.

In this section, I focus on the *deon*tological side of Esposito's communal contract. I examine various implicit references and dimensions in his analysis of the etymology of *communitas*. I situate this work within the broader problematic of wresting community from the trappings of the political economy of the proper.

Esposito begins by deconstructing the modern tradition in political philosophy. The modern tradition, he argues, writes community off by employing a "dubious homology" that confuses community with the *res publica* (5). This confusion raises three immediate problems. First, public is an excessively vague term. Second, the *res* connotes that the public is a thing with its own qualities. Third, the public always includes its antipode, the private. These three issues raise a whole series of aporias that confound modern notions of community. In his introduction, Esposito attempts to loosen the connection between community and the public so that he can begin to think about community beyond the tradition of political philosophy and its fidelity to the political economy of the proper.

Esposito's etymological analysis of the *munus* can be read as an attempt to readdress the classical deontological problem of *civic duty*. Esposito provides an alternative perspective on the "communitarian turn," which Gerard Delanty cites as the movement "'from contract to community'" (56).[6] His notion of communal duty represents a heterodox communitarian form of citizenship. It emphasizes duties owed to the community as opposed to the privileges, or rights, bestowed on individuals in the liberal tradition. The modern social contract, Esposito argues, undermines our "communal bond" when it institutes an "immunitarian" model of citizenship. Individuals are granted immunity from all forms of communal responsibilities.

The modern paradigm of immunity began when Hobbes used the social contract to institute a border between "the originary dimension of common living" and the modern "juridically 'privatistic' and logically 'privative' figure of the contract" (13). When he based our commonality in our "capacity to kill" and "the possibility of being killed" (26), Hobbes left us with a stark choice between remaining in a perpetual state of "terror" or converting this destructive element of our relations into a productive model based on "fear." By transposing "originary" and "uncertain" fear into "artificial" and "certain" fear, Hobbes essentially cut us off from our relationships with each other. Esposito calls this "absolute dissociation" in the social contract "the crime of community" (27). The social contract erected an artificial barrier that immunizes us from each other. "Subjects," he continues, "have nothing in common since everything is divided between 'mine' and 'yours': division without sharing [*condivisione*]" (28).

When communal duties are formulated within this private political economy, political philosophers are forced to deal with the issue of compensation (6) or *remuneration* (4, 139). Here, Esposito taps into a long tradition in the political economy of the proper. In *Politics*, for example, Aristotle provided a detailed examination of this problematic. In fact, his formulations still define the parameters that shape how contemporary political theorists such as Esposito broach the problem of remunerating civic duties. If we focus on the rule of the Multitude, leaving aside the rules of the One and the Few, there remain but three solutions.

The first calls for public duties to be performed without remuneration – a logically sound approach, but hampered by the problem of balance. Because the multitude is practically excluded from holding public office, this solution results in a politics dominated by the private interests of a small group of oligarchs. The second solution attempts to counterbalance the exclusionary effects of the first, by offering all political representatives remuneration. This model, too, falls short of its goal of being wholly inclusive because private economic interests still dominate politics. In both, a professional class of political oligarchs holds a monopoly over the political apparatuses who represent, directly or indirectly, the economic interests of the wealthy. Thus, the first two solutions remain trapped in the aporia of the private economy. Each calls for a strict, yet practically impossible, separation between private and public interests.

At stake in this discussion is the political conflation of private interests and public interest. This is a lesson found in Hannah Arendt's detailed analyses of "privation" (*Human Condition*) and "totalitarianism" (*Origins*), Jürgen Habermas's "refeudalization of the public sphere," C. Wright Mills' "power elite," and Robert Michels' "iron law of oligarchy." Not everyone is troubled by this conflation. Liberal Pluralists, for example, continue to preach Robert Dahl's position that competition between different interest groups is the driving force of modern democracies. Others, such as neo-Marxists, argue that we have exaggerated this conflation because modern politics are "relatively autonomous" from the economy (Poulantzas). Communitarians argue that the predominance of private economic interests in modern politics could be curtailed if properly infused with communitarian ethics. Here we find appeals to Alexis de Tocqueville's "civic volunteerism" or Max Weber's distinction between politicians who "live for politics" and those who "live from politics." Even radical democrats who claim to be beyond this reductive discourse, such as Ernesto Laclau and Chantal Mouffe (*Hegemony*), remain trapped within its logic. Neologisms appealing to apolitical, neo-political, or post-political are empty phrases whenever conceived within the discourse of the political economy of the proper. The core issue is how to move beyond the trap of distinguishing between private economic interests and political public interests.

The third solution is found in the orthodox versions of communism. When property is communalized, private property continues only in the diminutive state of personal possession. Gradually the distinction between public and private interests would disappear along with the question of remunerating public duties. But is this really a solution or is the private merely absorbed into the public in this model? Many read *community*, or *communism*, within the Augustinian model of *com-unus*. Here, individuals are completely absorbed by a collective subject, which in extreme cases is ascribed its own metaphysical, ontological qualities. Where the proper is appropriated by a community, the multitude is annihilated by the One. Whenever community is conceptualized as a "homogeneous totality," argues Esposito, community is subjected to the ethos of modern "nihilism" (*Communitas* 147). Totalitarianism is so inclusive that it closes off and contains the very division that is necessary to form communal relationships.

This is certainly not an exhaustive list, but I do not want to belabour the issue. What I want to show is how Esposito provides a subtle, yet important and immanent, critique of the third option. Whenever the common is configured within the auspices of the political economy of the proper, the common is represented as the bearer of the proper. A simple return to the original meaning of the *communitas*, Esposito contends, reveals that on the contrary the common is "what is *not* proper" because it "begins where the proper ends" (3). The proper is a derivative of the *proprius* (one's own, particular to itself), which is ultimately the private. The common is thus "improper."

The impropriety of the common is likewise translated into the public/private dichotomy. The public represents the quintessential "other." It is a "voiding, be it partial or whole, of property into its negative" because it removes "what is properly one's own [*depropriazione*]" by "invest[ing] and decenter[ing] the proprietary subject, forcing him [*sic*] to take leave of himself, to alter himself" (7). The public deprives subjects of their property. This does not mean that he is advocating for a simple reversal of the order of possession from private to public property. He actually argues that we have to stop thinking about the common as the public, partly because the public is too closely aligned with the private. The common is neither private nor public, he claims (Esposito, "Community").

Esposito's two-fold deontological solution begins with a re-evaluation of the role of *interest* in the traditional political economy of the proper. Within this economy interest is always treated in a proprietary sense as if it is a "compensation for loss" (*interesse*). One has an interest in community precisely because the more one gives, the more one can take back for oneself. Robert Putnam's communitarian notion of "social capital" falls within this self-interested model of community. The more one personally invests in one's community, the more social capital one accumulates, which can be cashed in for favours. Such a community works like a marketplace that people invest in to secure future returns in the form of favours. Reciprocity is thus formulated on the basis of self-interest.

Like Nancy (*Inoperative* 93), Esposito turns to the etymological construction of interest qua *inter-esse* in search for an opening.[7] The Hobbesian social contract, Esposito notes, sacrifices "not only the *inter* of *esse* but also the *esse* of *inter* in favor of individual interest" (*Communitas* 141). If we return to the ontological significance of the term *inter-esse*, then it might be possible to separate interest from the political economy of the proper. Being (*esse*), not property, interests us precisely because being is between (*inter*) us, as Nancy phrases *interesse* in "Compearance" ("La Comparution").[8] It is in this realm where the double sense of Esposito's deontological contract operates.

Esposito argues that a breach of the political economy of the proper is found in the three etymological senses of the *munus*: *onus* (load/burden), *officium* (office), and *donum* (gift) (*Communitas* 4). Each is characterized as a "duty" (*dovere*). Their deontological connotations are carried through in his subtle play on the contradictory senses in **mei-*, the root of the word *munus*: transformation/opening up (mutable, exchange); lessen (lack, diminish, minus); and binding/closing off (duty, obligation, indebtedness). That is, the *munus* is a deontological contract that opens up and transforms subjects (exchange), diminishes them to the point that they are wholly lacking (expropriation), and binds them to contractual obligations (indebtedness). It is a contract, we could say in a Nancean vein, that is written in the very ex of their common alteration. Its binding, or obligatory, characteristic is reinforced by the fact that each is commonly exposed to the lack, which Esposito argues is the common. It results in a two-fold deontological exchange that opens each to the binding obligation of performing services on behalf of community.

In reference to Marcel Mauss and Émile Benveniste, Esposito argues that the duty characteristic of the *munus* is presented in the form of a "gift." Were one to accept it, one would become obliged "to exchange it in terms of goods or service" (ibid.). Contrary to Mauss's notion of gift exchange and Putnam's notion of social capital, this gift exchange does not occur in a give-and-take form of reciprocity; rather, it is an "unrelenting" obligation where "one *must* give [...] because they *cannot not* give." It is also an "unequivocal" exchange that places the receiver in debt to, "'at the disposition of,'" "or more drastically 'at the mercy of' someone else" (5). When we are contracted together in the *munus*, the direct "correspondence" of giving and receiving is interrupted. It is, in short, an interruption of the dialectic of alienation and appropriation.

Esposito hints at this disruption, I believe, by making a play on the etymological connection of "debt" and "duty." Both words derive from *debere* (to owe), which breaks down as *de-habere* or de-having. On the *deon*tological end, debt/duty expropriates subjects. Subjects are placed in an impossible position of not having, holding, or possessing that which is owed. Their sense of having is negated: "'I owe *you* something,' but not 'you owe *me* something'" (6). In this de-having, the prospect of the returned favour is radically removed from the relationship. The communal marketplace is thus immediately displaced when each is opened up and exposed to the common. This is a different sense of the common and of debt than is found today in the West. Our debt crisis concerns private financiers who are empowered to repossess private property in exchange for money owed. Communal debt, however, expropriates more than property in the material sense. It also expropriates our "initial property," "of the most proper property, namely," our "very subjectivity" (7).

A deontological community isn't formulated in terms of "having" in the proprietary sense, but by a "debt" that contracts us together (com-) in the gift of the *munus*. Such an "exchange relationship" does not follow the proprietary contractual modalities of "give" (*dare*) and "take" (*prendere*) (4). We are drawn together into a "transitive act of giving" that has nothing to do with the "stability of a possession and even less the acquisitive dynamic of something earned, but loss, subtraction, transfer" (5). We are contractually obliged to give "something that one can *not* keep for oneself and over which, therefore, one is not completely master" (ibid.).

> The fact that the Romans understood by the term *munus* only the gift [*dono*] that was made and never the gift received (which was instead denoted in the word *donum*) signals that it is lacking in "remuneration," and that the breach of a subjective material that it determines remains as such, that is incapable of being made replete, made whole, or healed over; that its opening cannot be closed by any sort of compensation or reparation if it is to continue in fact to remain shared [*condivisa*]. The reason is that in the concept of "sharing with" [*condivisione*], the "with" [*con-*] is associated with dividing up [*divisione*]. (139)

One might question whether we should replicate a formula found in a Roman notion of community, an imperial power and slave society with a highly stratified notion of citizenship, governed by the Few. Further, the elite Roman citizens understood the de-having as a *tax* owed. Should community be predicated in these terms? Without further context, this theory could be read as another conservative rendition of communitarianism, moreover a romantic conservative version. This, however, would be an unfair conclusion; Esposito does clarify his political leanings in his discussion with Nancy (Esposito and Nancy, "Dialogue") and in the essay contained in this present issue (Esposito, "Community"). As with the other theorists in this literature, Esposito implores us to differentiate between the spirit of this notion of community, while discarding its form. I do believe, however, that Esposito could further elaborate on the relationship between his notion of *communitas* and the Roman notion beyond providing an etymological genealogy of the term. Nevertheless, the political connotations of his theory do become clearer in the second part of *Communitas*, which I will now discuss.

part iii: de*onto*logy and nihilism

> [W]hat the members of a community share [...] is rather an expropriation of their own essence, which isn't limited to their "having" but one that involves and affects their own "being subjects." Here the discourse follows a crease that moves from the more traditional terrain of anthropology or of political philosophy to that more radical terrain of ontology: that the community isn't joined to an addition but to a subtraction of subjectivity, by which I mean that its members are no longer identical with themselves but are constitutively exposed to a propensity that forces them to open their own individual boundaries in order to appear as what is "outside" themselves. (*Communitas* 138)

In the following section, I examine how Esposito complements his *deon*tological communal contract outlined in the first half of *Communitas* with an existential de*onto*logical model of a communal contract in the second half of this book. This work is set up in the last two chapters on Heidegger (86–11) and Bataille (112–34), but it only comes to fruition in the appendix "Nihilism and Community" (135–49). This trajectory is important for examining how he has contributed to the literature that seeks to conceptualize community in an ontological fashion.

Esposito's existential strain is grounded in Heidegger's writings on the event of existence, either as the "Call of Being" (*Being and Time*) in his earlier works or as the "*Ereignis*" in his later *Ereignisdenken* ("Time and Being"; "Principle of Identity"). For Heidegger this event opens us up and exposes us to Being.

Esposito draws from Heidegger's claim that modern humanity remains closed to Being precisely because we treat Being as something that we can *have*. Being is not something that we can identify, hypostatize, and thus appropriate. The *Ereignis* (propering event) temporarily suspends our appropriative grasps so that being is prioritized over having.

In *Communitas*, Esposito repeatedly criticizes the "affirmative entification" of community in contemporary theories. His *munus* performs the function of de-having (debt/duty – *debere*). He differs from Heidegger, however, by trying to rethink this de-having in a plural fashion. Here he follows Nancy's efforts to pluralize the singular outlook of Heideggerian existentialism. For Nancy and Esposito, the event of existence must be read within a plural register as *being-with*. For Nancy, Being is always experienced in the plural qua beings or qua existents (see Nancy, "La Comparution"; "Decision of Existence"; "Of Being Singular Plural"; "Surprise of the Event").[9]

Nancy's revision of this problematic largely concerns the technical problems of rethinking Being as a plural phenomen*a* of beings; Esposito, however, carries this problematic into a contentious terrain in Heideggerian scholarship. His two-fold deontology forces him to deal with the "identity of ethics and ontology" (*Communitas* 90). Esposito is neither ostentatious nor brazen here. He disputes Levinas's charge that Heidegger "sacrifices the first to the second" and Ricoeur's claim that Heidegger leaves ontology "'on the threshold of ethics.'" He is aware that to venture down this path with Heidegger with a communitarian motive in hand is dangerous. He carefully places some distance between himself and Heidegger by engaging with Heidegger's writings on being-with, belonging, and origination in a critical tone. He then follows with a chapter on Bataille's reconfiguration of Heideggerian existentialism in a communal fashion (112–34). This recourse to Bataille gives Esposito just enough distance to reconfigure Heideggerian existentialism in "Community and Nihilism."[10]

In "Ecstasy," Esposito reiterates the "communitarian" leanings in Heidegger's earlier writings (92–97), then he questions Heidegger's uncritical reference to the "*Volk*" and "destiny" (97–102), and he finishes by challenging Heidegger's misreading of the place of "origin" in Hölderlin's work (102–11). According to Esposito, we need to turn to Bataille to right the problem of "the origin and destiny of community" – the subtitle of *Communitas*. Esposito points to a pair of steps Bataille takes that help us reconsider how deontology can be understood in a two-fold fashion.

First, unlike Heidegger's formulation of being-with, Bataille places the *cum* at the end. For Bataille, community is experienced through the horizon of death, which represents the "nullification of every possibility in the expropriating and expropriated dimension of the impossible" for "death is our *common* impossibility of being what we endeavor to remain, namely isolated individuals" (121). Esposito reads Heidegger's notion of death as one's most proper (*Eigentlichkeit*) possibility literally, so he turns to Bataille for an alternative account. With Bataille, the other's death, or even an act where two people risk their lives, opens us up, exteriorizes us, and places us in proper communication with each other. This common exposure disrupts our appropriative capacities just enough for communication to occur without pretension. The experience of a "'common nothingness,'" claims Esposito – in a manner similar to Alexandre Kojève's synthesis of Hegel, Heidegger, and Marx – exposes us to our common and shared incapacity to appropriate what is properly our own: our finitude. Bataille thus figures as a "radical anti-Hobbesian" because his model prioritizes openness and exposure, which are communitarian impulses, as opposed to immunitarian closures and barriers (123–24).[11]

There is also a *deon*tological strain in Bataille's critique of the "restrictive economy" of the Hobbesian social contract tradition (*Communitas* 124). In its place, Bataille "refers to a 'munificence' purged of any mercantile remnants" (ibid.). Notably, munificence breaks down etymologically as *munus-facere*: to do or perform a *munus*. For Bataille, the "gift," qua sacrifice, to the community is "life." Here Esposito treads back into the tradition that connects the *munus* to the logic of sacrifice. Altruism is traditionally enacted through a sacred duty of giving oneself over as a gift that is owed to *communitas* without any expectation of being remunerated. This is clearly an expropriative act; yet it also resonates with the Hobbesian paradigm of sacrifice.

The relationship between community, contractualism, and existentialism is drawn from Nancy's rejoinder to Bataille in his essay on the "unsacrificeable" (Nancy, "Unsacrificeable"). Nancy argues: "'finitude' means that existence can't be sacrificed" (cited in *Communitas* 128). Existence is itself an offering that is, in Esposito's paraphrase, "more originary than every sacrificial scene, but offered to nothing and to no one and therefore not sacrificed" (ibid.). Esposito argues that if the sacrificial logic is replaced by his reading of nihilism, then the "Hobbesian moment" can be surpassed.

Esposito's nihilistic reading of munificence brings together community, contractualism, and existentialism. In his contract, de-having leaves each of us exposed to nothing. This ontological opening unto being exposes us so thoroughly that we are left in a state of incapacitation. Each is forced to face that which one can never be exonerated or immune from and that which one can never properly appropriate: their finitude. "Finite subjects" are "cut by a limit that cannot be interiorized because it constitutes precisely their 'outside'; the exteriority that they overlook and enters into them into their common non-belonging" (7).

This side of his argument resonates with Nancy's Heideggerian account of the exposure that occurs when we "compear," that is, appear together (Nancy, "La Comparution"). For Esposito, exposure to finitude interrupts our appropriative impulses. Our need to appropriate others, ourselves, or common items, is extensively disrupted. When we are exposed, we experience "a dizziness, a syncope, a spasm in the continuity of the subject" (*Communitas* 7). Compearance occurs in an ontological opening that is always a plural, and, thus, communal event. Community occurs, claims Esposito, in being exposed to one's "radical impropriety." Community is this "hole into which the common thing continually risks falling" (8). It is this risk of "falling" into a "sort of landslide" that constitutes our danger of living together, our finitude, which threatens to penetrate our artificial borders and turn each of us inside-out. Being is always experienced qua being-with. It is this lack, the very lack in our being, which is "configured as an onus" (6). We share this lack that divides us.

Esposito's celebration of nihilism is predicated in a non-romantic register. To think about community "in a way that is able to meet the needs of our own time," which is "characterized by a fully realized nihilism," then the relationship between community and nihilism must be readdressed (136). The point where "they cross each other," he contends, is the "no-thing." The *munus* can only take place in the "radical terrain of ontology" (138) where members of community are subtracted, and exposed. Identifiable boundaries are forced open such that each appears "as what is 'outside' themselves" (ibid.). Each becomes an "'other'" in "a chain of alterations that cannot ever be fixed in a new identity" (ibid.). Esposito's altruistic notion of community is thoroughly defined by alterity or otherness, not by sameness and identity. This procession is not enacted by the other, such as Levinas's second person philosophy, or by a sacrificial gesture, but by the *munus*, which in *Third Person* he calls the "impersonal third."

Community, Esposito argues, is the "*esse* as *inter*, not the relation that shapes being [*essere*] but being itself as the relation" (*Communitas* 139). It is the "interval of difference, that spacing that brings us into relation with others in a common non-belonging, in this loss of what is proper that never adds up to a common 'good'" (ibid.). By prioritizing difference over sameness, spacing over place, exposure over safety, risk over security, etc., he not only reverses the traditional characteristics of community but also makes a play on the division/sharing/taking leave that is found in the notion of *condivisione*, like Nancy's *partage*. That is, to be exposed in the division and sharing is to depart from the traditional immunitarian community. Community signifies nothing more than the relationships that occur in a risky situation, in a flight into insecurity, which cannot produce the traditionally understood "effects of commonality, of association, and of communion" (140). We are not immune to others in this situation. If we refuse to see community in this manner, he contends, then we will continue to find ways to "contain the dangers," "immunize" ourselves, and cut ourselves off from community, such as Hobbes does in his contract (141). We will continue to experience community as either a division without sharing (the immunitarian social contract) or a sharing without division (absorption into a grand communal individual).

Fittingly, Esposito ends by turning to Heidegger first, then Bataille, and finally Nancy. The first two are "necessary" references in the "new philosophical reflection about community" in Italy and France by Agamben, Blanchot, Nancy, and of course, himself (see 148 fn. 13). It will suffice to end this section on the lessons he draws from this theoretical trajectory.

Heidegger demonstrated that relationships only occur across nothingness. If community were something or, worse, everything, then it would represent an annihilation of all relationships. It would result in – recalling the twofold appropriation discussed above – a sheer implosion of unity. Modern nihilism actually abolishes the distance that is necessary for relationships to occur. Communitas, on the contrary, is a "unity *in* distance and *of* distance" (145).

Bataille devised a more concrete way to conceptualize the relationship between community and the no-thing. For him, "nihilism isn't the flight of meaning or flight from meaning, but rather meaning's enclosure within a homogeneous and complete conception of being" (146). Nihilism is a complete enclosure, immanentism, that excludes alterity and difference. Nihilism, in other words, is a product of the political economy of the proper which directs us to appropriate difference. Difference is thus reduced to sameness, which is precisely how modern notions of community operate.

Finally, Nancy, "more than anyone else has the merit of clearing a way forward in the closed thought of community" (148). Nancy doesn't necessarily rejoice in the destructive aspect of nihilism, but he does force us to take issue with the "end of every generalization of sense" and the "emergence of singular meaning that coincides with the absence of meaning" (149). It is an opening, or a clearing, to the "nothing held in common that is the world that joins us" (ibid.). Singular plurality, Esposito claims, represents a new path for reconceptualizing community beyond the semantics of the proper and in an existential fashion.

conclusion

Earlier I raised Balibar's critique of those who deconstruct the proper. If addressed to Esposito's notion of *communitas*, the question becomes: doesn't his theory of community lead to an "*abyssal alienation*" (Balibar, "'Possessive Individualism'" 315)? There are two sides to this term. On the one hand, there is the problem of the abyss. Although Esposito's community lacks an identifiable and substantial foundation, basis, or ground upon and through which we relate with one another, this doesn't mean it lacks a nexus. The nexus for him is the *munus*, which draws people together in relationships of duty and obligations. I want to leave aside any further discussion about the "bottomless" aspect of Esposito's theory. For the remainder of this paper, I focus on the problem of alienation, which is an issue that critics have repeatedly raised against Agamben, Esposito, and Nancy.

Balibar's question is loaded. It is framed as a normative defence of the political economy of the proper. Despite all I said above, especially in the section on the dialectic of alienation and appropriation, this question requires further exploration. If Esposito has not advanced community beyond alienation then it remains trapped in the traditional communitarian model wherein community – even if it is nothing, thus not a "hypertrophic" thing that engulfs individuals – still takes absolute priority.

Community is neither a property nor a proprietor, but the "improper," which he clearly understands in the more limited sense of individual property or one's own. We are indebted for the gift of community. The exchangeable good, or the "'tribute' that one pays in an obligatory form" (*Communitas* 5), is a community service that contributes to the common good. Esposito argues that our indebtedness to community radically dispossess us. In this contract the commercial currency of the political economy of the proper is exchanged and cashed into an ethical political economy. In Esposito's work, this is signified by his reconfiguration of the proprietary reading in the traditional individual vs. community dichotomy with a two-fold deontological formulation of *immunitas* vs. *communitas*. Duty and Being are factored into this distinction.

Even if Esposito's secularized reading of *munificence* syncopates the traditional models of the sacrificial community, there remains an altruistic basis at the core of his thought. This basis opens him to the charge of radical alienation, even if he does not conceptualize it as traditional political economists have. A brief comparison between his theory of *communitas* and Anglo-American communitarian theory can help to clarify this issue.

In *Communitas* Esposito situates his thinking beyond the current impasse between communitarians and cosmopolitans in Anglo-American theory. Alongside communitarians he prioritizes civic responsibilities over cosmopolitan civil rights. Alongside cosmopolitans he is skeptical about communitarian "micro-communities" predicated on "ethnic, linguistic, and cultural identity" because they produce closed and exclusive communities (54–55). He does not call for an abstract and universal notion of a world community or a global citizenry; yet he questions why our models of community prioritize place, home, and proximity. It is in this point where he combines contemporary calls for community to occur across space, not place, with Nancy's notion that community occurs as a spacing that extensively, yet ephemerally, erases place. Space is empty. It cannot be subjected to the logic of identification and appropriation. In *communitas*, one is so thoroughly dis-placed, dis-positioned, and exposed that one is made uncomfortable, insecure, and thus homeless. Community, in other words, must be reconceptualized in a fashion that is more adaptable to the new global world that we share. All place-based models of closed and exclusive communities – cultural, ethnic, or national – must be disrupted.

The charge of alienation, I believe, should not be directed at his call for a radical model of altruism, which is based on a notion of the *munus* that includes everyone, but at the ramifications of his prescriptions. Like other communitarians, Esposito prioritizes duties over rights – not just because duties ensure that members of a community act as responsible citizens but also because he is critical of the immunitarian logic that results from the relationship between civil rights and the *proprius* (see also Esposito, *Third Person*; "The *Dispositif*"). Too much reliance on rights, in other words, erodes the communal fabric of social relationships. Esposito does not mourn the loss of a mythological familial order in the face of an all-intrusive "Nanny" state, like Amitai Etzioni; rather, he comes close to Arendt's communitarian notion of civic republicanism. Like Arendt, Esposito is wary of the proprietary logic that underscores identity politics, liberalism, and mass privation. A problem arises, however, when we turn to the content of duty.

Esposito's notion of duty remains too open and indefinite. In a sense, it operates as an empty signifier that can be translated into many different contexts. This helps to keep the term open and inclusive. Yet there remains the crucial problem of translation and applicability in this theory. Either we hold fast to the terrain of duty ethics or we translate obligation into the realm of morality. The former leads to a cold, neo-functionalist model of community, while the latter becomes intertwined with parochial moral vernaculars that immunize locals from outsiders. The latter represents a hyperbolic sense of the proper which extends from identity claims to moral distinctions between proper and improper conduct. Esposito clearly wants to avoid this terrain, which leaves his theory in the other. It is here that the charge of alienation could rest.

As it stands in *Communitas*, his theory requires a stronger infusion of the social. This doesn't mean that he must revert to a commericialized notion like Putnam's "social capital" cited above, but there needs to be a stronger sense of social relationships. He does mention "social ties" in his essay "Community, Immunity, Biopolitics," but more work could be done on this end. Absent the social, we end up with a dry, functionalist, duty-bound notion of community at best or a severe, repressive, duty-bound notion of community at worst. I doubt this is his intention, as he is clearly writing from within the strong tradition of leftist Italian theory, but it is a common problem we find in many political thinkers who tend to ignore the social. Civic duties, engagements, and contributions remain socially alienating (not necessarily economically alienating in Marxist terminology) without a sense of the social.

Esposito's theory is not closed off, or even opposed to sociality, it is just missing. But the question remains: how can thoroughly immunized individuals be compelled to conduct their communal duties? An ethics of ontology is absolutely necessary given that we are now living in a period of global warming, but thus far most immunized individuals remain resistant to serious communal engagement. Since the barriers are so strong – capitalism, neo-liberalism, neo-colonialism, privatization, de-regulation, class polarization, racism, sexism, etc. – we need more ammunition to shake ourselves out of our current immunitarian malaise.

If immunized individuals are averse to the *munus*, and many more are excluded from it, then we have to be more inventive in finding ways to encourage people to act on behalf of the gift of community. If we do not want to placate to self-interests and if the altruistic measure of moral compulsion, shame, is now impotent (Tönnies), what options do we have at our disposal? People living in advanced capitalist societies are less likely to voluntarily join communal associations (Putnam). Moreover, communal duties can be carried out in an automatic, indifferent, even careless manner. Such acts would be communal in form only. If immunized individuals are to adhere to the deontological commandment to pay tribute for the gift of community, retribution cannot appear like a punishment paid for a crime committed. Esposito is right to point out that our current ontological predicament requires a turn to altruism, but the socially alienating aspects of the *munus* have to be offset by a more inviting and welcoming sense of belonging. Without a social tie, in short, *communitas* is probably too taxing for immunized individuals.

There is an opening in his thought to bring forward a social dimension: *sharing*. An open notion of sharing could help to foster a warmer and inviting sense of belonging. I am not suggesting he turns to the anachronisms of love, friendship, or family to project the social into this model, but without a social atmosphere – I am not saying spirit – of sharing, *communitas* could be read as a cold, functional ethical machine, not unlike the ascetic communities existing on the fringes of modern societies. The basis of their communal bond is alienation; sociality, in short, could rectify this issue in this theory.

notes

I would like to thank Kathy Bischoping, Kristin Hole, Anne O'Byrne, and Jon Short for providing thoughtful feedback on an earlier draft of this paper. Thank you also to the anonymous peer-reviewers for constructive feedback and to the editorial team at *Angelaki*.

1 For a brief overview of this topic see C.B. Macpherson's work on the "possessive individual" (*Political Theory*; "Liberalism"; and "Meaning"), Vandana Shiva's critique of the Western property prejudice, and Étienne Balibar's critique of Heideggerian efforts to deconstruct this political economy ("'Possessive Individualism'"). For an examination of the relationship between Lockean political

economy and colonialism see Armitage; and Pateman.

2 This is not to say that political theology does not play an important part in Esposito's analyses of *communitas* or its constitutive other, *immunitas*. A full chapter called "The *katechon*" is dedicated to political theology in *Immunitas* and it is addressed in the opening pages of *Communitas*. This theme is also present in Eposito's writings on the impersonal, such as *Third Person*, "For a Philosophy of the Impersonal," and "The *Dispositif* of the Person." For further commentary see Campbell; Goodrich; and Barkan.

3 For an excellent discussion of Esposito's use of "im-" in his writings on the "impolitical" as both a position internal to politics and against politics see Bosteels. Esposito likewise uses this double sense when writing about the improper. It is a spacing, a gap, even a hole inscribed within the very political economy of the proper that is essential to its core functions. To appropriate and close it off is to put an end to politics, which, Esposito claims, results in "totalitarianism" ("Totalitarianism"). Esposito seeks to exploit this space to circumvent and traverse it. He employs this deconstructive motif across most of his writings.

4 See Esposito, "Community, Immunity, Biopolitics."

5 See both Bosteels and Neyrat for critiques of Esposito on this score; see both Norris and Elliott for similar critiques of Nancy.

6 Gerard Delanty is referring to a collection of essays on communitarianism edited by Fred Dallmayr in 1978.

7 The first footnote of *Communitas* praises Nancy's *The Inoperative Community*, which Esposito claims in an allegorical fashion is "a text to which I owe an unpayable debt, as is the case for every *munus* given us in the form of the most unexpected gift" (*Communitas* 1 fn. 1). The full significance of these words should be apparent from the examination in this section. It is also noteworthy that the last footnote of the work is dedicated to Nancy's *The Sense of the World* (*Communitas* 148 fn. 14).

8 "Finite existence is necessarily divided/shared (*partagée*). 'Politics' must designate what *interests* us in the 'common'. The stakes are the interest (that which matters) of *inter-esse* (simultaneously: 'to be between', 'to be separated', 'to differ', 'to be amongst', 'to participate')" ("La Comparution" 94).

9 The Heideggerian strains in this debate have been well documented. See Bird; Bernasconi; Critchley; Gratton; James; O'Byrne; Raffoul.

10 Nancy remains reluctant to treat community, even compearance, in an ethical formulation; however, for Esposito ontology is ethical (see Esposito and Nancy). Nancy focuses on the compulsion and force of the event itself, while Esposito employs a *deon*tological formula.

11 Of course, Bataille serves as the third person in the debate between Blanchot (*Unavowable Community*) and Nancy (*Inoperative Community*) (see Bird; Bernasconi; and Critchley). Esposito's placement of this chapter at the end of this book is significant in this regard, which he clearly signifies by placing a reference to Blanchot's *Unavowable Community* in the first footnote of this chapter.

bibliography

Agamben, Giorgio. *The Coming Community*. Minneapolis: U of Minnesota P, 1993. Print.

Arendt, Hannah. *The Human Condition*. Chicago: U of Chicago P, 1998. Print.

Arendt, Hannah. *The Origins of Totalitarianism*. New York: Harcourt, 1976. Print.

Aristotle. *Politics*. Trans. C.D.C. Reeve. Indianapolis: Hackett, 1998. Print.

Armitage, David. "John Locke, Carolina, and *The Two Treatises of Government*." *Political Theory* 32.5 (2004): 602–27. Print.

Balibar, Étienne. "'Possessive Individualism' Reversed: From Locke to Derrida." *Constellations* 9.3 (2002): 299–317. Print.

Balibar, Étienne. *We, the People of Europe? Reflections on Transnational Citizenship*. Trans. James Swenson. Princeton: Princeton UP, 2004. Print.

Barkan, Joshua. "Roberto Esposito's Political Biology and Corporate Forms of Life." *Law, Culture and the Humanities* 8.1 (2012): 84–104. Print.

Bernasconi, Robert. "On Deconstructing Nostalgia for Community within the West: The Debate between Nancy and Blanchot." *Research in Phenomenology* 23.1 (1993): 3–21. Print.

Bird, Greg. "Community beyond Hypostasis: Nancy Responds to Blanchot." *Angelaki* 13.1 (2008): 3–26. Print.

Blanchot, Maurice. *The Unavowable Community*. Trans. Pierre Joris. Barrytown: Station Hill, 1988. Print.

Bosteels, Bruno. "Politics, Infrapolitics, and the Impolitical: Notes on the Thought of Roberto Esposito and Alberto Moreiras." *New Centennial Review* 10.2 (2010): 205–38. Print.

Campbell, Timothy. "'Enough of a Self': Esposito's Impersonal Biopolitics." *Law, Culture and the Humanities* 8.1 (2012): 31–46. Print.

Critchley, Simon. "With Being-With? Notes on Jean-Luc Nancy's Rewriting of Being and Time." *Studies in Practical Philosophy* 1.1 (1999): 53–67. Print.

Dahl, Robert A. *Who Governs? Democracy and Power in an American City*. New Haven: Yale UP, 1961. Print.

Delanty, Gerard. *Community*. 2nd ed. New York: Routledge, 2010. Print.

Derrida, Jacques. *Specters of Marx: The State of the Debt, the Work of Mourning, and the New International*. Trans. Peggy Kamuf. New York: Routledge, 1994. Print.

De Tocqueville, Alexis. *Democracy in America*. Vols. 1 and 2. Trans. Henry Reeve. Toronto: Bantam, 2000. Print.

Elliott, Brian. "Community and Resistance in Heidegger, Nancy, and Agamben." *Philosophy and Social Criticism* 37.3 (2011): 259–71. Print.

Esposito, Roberto. *Bíos: Biopolitics and Philosophy*. Trans. Timothy C. Campbell. Minneapolis: U of Minnesota P, 2008. Print.

Esposito, Roberto. *Communitas: The Origin and Destiny of Community*. Trans. Timothy C. Campbell. Stanford: Stanford UP, 2010. Print.

Esposito, Roberto. "Community, Immunity, Biopolitics." Trans. Zakiya Hanafi. *Angelaki* 18.3 (2013): 83–90. Print.

Esposito, Roberto. "The *Dispositif* of the Person." *Law, Culture and the Humanities* 8.1 (2012): 17–30. Print.

Esposito, Roberto. "For a Philosophy of the Impersonal." Trans. Timothy Campbell. *New Centennial Review* 10.2 (2010): 121–34. Print.

Esposito, Roberto. *Immunitas: The Protection and Negation of Life*. Trans. Zakiya Hanafi. Cambridge: Polity, 2011. Print.

Esposito, Roberto. *Third Person: Politics of Life and Philosophy of the Impersonal*. Trans. Zakiya Hanafi. Cambridge: Polity, 2012. Print.

Esposito, Roberto. "Totalitarianism or Biopolitics? Concerning a Philosophical Interpretation of the Twentieth Century." Trans. Timothy Campbell. *Critical Inquiry* 34.4 (2008): 633–44. Print.

Esposito, Roberto, and Jean-Luc Nancy. "Dialogue on the Philosophy to Come." Trans. Timothy Campbell. *Minnesota Review* 75 (2010): 71–88. Print.

Etzioni, Amitai. *The Spirit of Community: The Reinvention of American Society*. New York: Simon, 1994. Print.

Goodrich, Peter. "The Theatre of Emblems: On the Optical Apparatus and the Investiture of Persons." *Law, Culture and the Humanities* 8.1 (2012): 47–67. Print.

Gratton, Peter. "Mitsein and Sharing: Heidegger and Nancy on the Political Community." *Review Journal of Political Philosophy* 2.1–2 (2004): 33–58. Print.

Habermas, Jürgen. *The Structural Transformation of the Public Sphere: An Inquiry into a Category of Bourgeois Society*. Trans. Thomas Burger. Cambridge, MA: MIT P, 1989. Print.

Heidegger, Martin. *Being and Time*. Trans. John Macquarrie and Edward Robinson. New York: Harper, 1962. Print.

Heidegger, Martin. "The Principle of Identity." *Identity and Difference*. Trans. Joan Stambaugh. Chicago: U of Chicago P, 2002. 21–41. Print.

Heidegger, Martin. "Time and Being." *On Time and Being*. Trans. Joan Stambaugh. New York: Harper, 1972. Print.

James, Ian. *The Fragmentary Demand: An Introduction to the Philosophy of Jean-Luc Nancy*. Stanford: Stanford UP, 2006. Print.

Kojève, Alexandre. "In Place of an Introduction." *Introduction to the Reading of Hegel: Lectures on the Phenomenology of Spirit*. Ed. Allan Bloom. Trans. James H. Nichols Jr. Ithaca, NY: Cornell UP, 1980. 3–30. Print.

Laclau, Ernesto, and Chantal Mouffe. *Hegemony and Socialist Strategy: Towards a Radical Democratic Politics.* London: Verso, 1985. Print.

Locke, John. *Second Treatise on Government.* Ed. C. B. Macpherson. Indianapolis: Hackett, 1980. Print.

Macpherson, C.B. "Liberalism and the Political Theory of Property." *Domination.* Ed. Alkis Kontos. Toronto: U of Toronto P, 1975. 89–100. Print.

Macpherson, C.B. "The Meaning of Property." *Property: Mainstream and Critical Positions.* Ed. C.B. Macpherson. Toronto: U of Toronto P, 1978. 1–13. Print.

Macpherson, C.B. *The Political Theory of Possessive Individualism: Hobbes to Locke.* London: Oxford UP, 1962. Print.

Max, Karl, and Frederick Engels. "The Communist Manifesto." *The Economic and Philosophical Manuscripts of 1844.* Trans. Martin Milligan. Amherst, NY: Prometheus. 1998. 203–43. Print.

Marx, Karl, and Frederick Engels. *The German Ideology.* Amherst, NY: Prometheus, 1998. Print.

Michels, Robert. *Political Parties: A Sociological Study of the Oligarchical Tendencies of Modern Democracy.* Trans. Eden Paul and Cedar Paul. New York: Free, 1962. Print.

Mills, C. Wright. *The Power Elite.* New York: Oxford UP, 1956. Print.

Nancy, Jean-Luc. "La Comparution." *La Comparution. Politique à venir.* By Jean-Luc Nancy and Jean-Christophe Bailly. Paris: Bourgois, 1991. 47–100. Print.

Nancy, Jean-Luc. "The Decision of Existence." *The Birth to Presence.* Ed. and trans. Brian Holmes. Stanford: Stanford UP, 1993. 82–109. Print.

Nancy, Jean-Luc. *The Inoperative Community.* Trans. Peter Connor. Minneapolis: U of Minnesota P, 1991. Print.

Nancy, Jean-Luc. "Of Being Singular Plural." *Being Singular Plural.* Trans. Robert D. Richardson and Anne E. O'Bryne. Stanford: Stanford UP, 2000. 1–99. Print.

Nancy, Jean-Luc. *The Sense of the World.* Trans. Jeffrey S. Librett. Minneapolis: U of Minnesota P, 1997. Print.

Nancy, Jean-Luc. "The Surprise of the Event." *Being Singular Plural.* Trans. Robert D. Richardson and Anne E. O'Bryne. Stanford: Stanford UP, 2000. 159–76. Print.

Nancy, Jean-Luc. "The Unsacrificeable." *A Finite Thinking.* Ed. Simon Sparks. Stanford: Stanford UP, 2003. 51–77. Print.

Neyrat, Frédéric. "The Birth of Immunopolitics." Trans. Arne De Boever. *Parrhesia* 10 (2010): 31–38. Print.

Norris, Andrew. "Jean-Luc Nancy on the Political after Heidegger and Schmitt." *Philosophy and Social Criticism* 37.8 (2011): 899–913. Print.

O'Byrne, Anne. *Natality and Finitude.* Indianapolis: Indiana UP, 2010. Print.

Pateman, Carole. "The Settler Contract." *Contract and Domination.* By Carole Pateman and Charles W. Mills. Cambridge: Polity, 2007. 35–78. Print.

Poulantzas, Nicos. *Political Power and Social Classes.* London: New Left, 1978. Print.

Putnam, Robert. *Bowling Alone.* New York: Simon, 1999. Print.

Proudhon, Pierre-Joseph. *What is Property?* Trans. Donald R. Kelly and Bonnie G. Smith. New York: Cambridge UP, 2005. Print.

Raffoul, François. "The Logic of the With: On Nancy's *Être singulier pluriel.*" *Studies in Practical Philosophy* 1.1 (1999): 36–52. Print.

Sartre, Jean-Paul. *Critique of Dialectical Reason.* Vol. 1. Trans. Alan Sheridan-Smith. New York: Verso, 1991. Print.

Shiva, Vandana. *Biopiracy: The Plunder of Nature and Knowledge.* Boston: South End, 1997. Print.

Tönnies, F. *Community and Society.* Trans. C.P. Loomis. New York: Harper, 1957. Print.

Weber, Max. "Politics as a Vocation." *Max Weber: The Vocation Lectures.* Ed. David Owen and Tracey B. Strong. Trans. Rodney Livingstone. Indianapolis: Hackett, 2004. 32–96. Print.

The front is a skin, a selective "membrane", admitting only the homogeneous, the assimilable, or rather, that which is heterogeneous but considered to be "beneficial."

Derrida, Echographies

If anything, the immune system must be interpreted as an internal resonance chamber, like the diaphragm through which difference, as such, traverses us.

Esposito, Immunitas *18*

No community is possible that would not cultivate its own auto-immunity.

Derrida, "Faith and Knowledge" 51

penelope deutscher

THE MEMBRANE AND THE DIAPHRAGM

derrida and esposito on immunity, community, and birth

For Esposito, the other to whom we are originally open is not sexually differentiated, whereas sexual difference is a frequent figure of alterity in Derrida's work. As they reflect on the history of political philosophy, it is true that Derrida and Esposito have both focused on the fraternity of political bonds. When Esposito asks, "what precisely is a 'fraternal' democracy?"(*Bíos* 173), the answer leads him, as it also does Derrida in *Politics of Friendship*, to an interpretation of fraternity as a mode of immunity. But the question will lead Derrida to closer interrogations of the absence of women and sisters in the tradition of political philosophy (*Politics* 149, 155, 186). It leads him to an interest in questions such as: what happens when the sister is made a case of the brother?; and: what is a friend in the feminine? (ibid. viii, 56); and to a deconstruction of the simultaneously non-literal and strangely literal status of the "brother" (ibid. 93) which permits the historical exclusion of the "sister" from political community and political claims. These are welcome interrogations from Derrida which Esposito does not broach.

Instead, the interest in fraternity will lead Esposito to a reflection more specifically on the preoccupation with birth as a biopoliticization of fraternity which becomes an increasingly violent approach to reproduction with the capacity to culminate in the thanatopolitical theatres of blood associated with auto-immunity (*Bíos* 169ff.). Interestingly, women do find themselves playing an important rhetorical role in Esposito's work at this point. This occurs in his conceptualization of both immunity and community, even though neither of these becomes the context for a philosophy inflected by sexual difference. He identifies specific forms of violence towards women – specific forms of thanatopolitics – at the extremes of the immune paradigm. And he draws on the figure of the mother, and her relationship to the fetus, to offer new images of community.

On the one hand, a feminist reading can welcome the critical attention Esposito gives to these thanatopolitical modes. On the other, the use of maternity as a positive figure of community is less promising. It arises in part because Esposito, identifying the thanatopolitical captures of reproductive life, proposes conceptual alternatives: just as new and transformative norms of life would replace normativized life; and just as a Merleau-Pontyian flesh is proposed by Esposito as a conceptual alternative to body; similarly, birth as exteriorization is proposed as an alternative to birth's thanatopolitical modes (of which the example proposed is the Nazi immune measure of – as Esposito describes it – suppressing birth (*Bíos* 169)).

But in the absence of any other significant role for sexual difference, any other means of distinguishing men and women, and any other specific reference to women in Esposito's work on the immune paradigm, women's role in his work is reduced to maternity – and this very much in contrast to the extensive status and complex treatment of sexual difference in Derrida's work.

One can, moreover, criticize not only the reduction of women to the role of maternity but also their reduction to a particular image of maternity. Consider, as an example, how the reconfiguration of maternity as an affirmative opening to life will lead Esposito, in his discussion of the immune paradigm in *Bíos*, to refer to Rwandan mothers who end up embracing the children who ensued from genocidal rapes: "that all Rwandan mothers of the war, when asked about their experience, declared their love for their children born of hate signifies that the force of life prevails once again over that of death" (*Bíos* 7). True, Esposito is engaged in a project of offering positive alternative images to the thanatopolitical capture of birth. As a result, however, where birth is associated (in this reconfiguration) with the figure of welcomed, unpredictable alterity of life, there is no *affirmative* representation of the fact that the multiple variations of anticipation of life must also include an anticipation of the monstrous.

Comparing Esposito's and Derrida's work more specifically from the perspective of their treatment of maternity allows a consideration of Derrida's indirect use of birth as figure of the monstrous *arrivant*, particularly in light of the fact that an affirmation of this inevitable possibility is missing from Esposito's work. I will argue that from the perspective of a Derridean reading, Esposito's work remains, *at least to this extent*, subordinated to the immune paradigm it also explicates, a persisting immune paradigm for which the representations of maternity are emblematic.

I figures of perversion

A number of commentators have begun to consider the ways in which Derrida and Esposito can be compared, but must also be differentiated, in their articulation of immunity.[1] Because these accounts diverge significantly, particularly in their conclusions and in the connotation each gives to the figure of auto-immunity, the similarities are, if anything, the more surprising, for on a number of points they do come into close proximity.

Though Esposito's work does not share the deconstructive form of reading that Derrida brings to the history of philosophy, linguistics, and contemporary politics, nor the specific interest in the differing and deferring work of *différance*, Esposito and Derrida agree on much concerning the incoherence, and dangers, of projections of immunization against alterity. According to both philosophers, such projects attempt the protection and defense of an ideal state which is represented as anterior – but is incoherent as such. For both, what is defended is in fact impossible, it cannot be immunized, kept "*sain et sauf*" ("Faith and Knowledge" 2), despite the efforts of the immunitary project. For Derrida any seemingly original state of ideal origin, identity, nature or community would be riven by *différance*, whereas for Esposito the most original state would always be the original expropriation of *munus*. And for both, immunitary projects belie their own impossibility and easily turn to violence. In *Immunitas*, immunity is introduced by Esposito partly as an "interpretive category" for phenomena as various as those discussed by Derrida: "phenomena such as the battle against a new resurgence of an epidemic, opposition to an extradition request for a foreign head of state accused of violating human rights, the strengthening of barriers in the fight against illegal immigration, and strategies for neutralizing the latest computer virus [...] – not to mention a terrorist attack" (*Immunitas* 1–2; see also *Terms of the Political* 132).

While both of them turn their attention to immunitary defenses which attempt an impossible and violent defense of an absent presence, the reasons for this impossibility are differently parsed. For Derrida, attempts at immunity, at keeping ideals "*sain et sauf*," do lead to many of the same violent extremes identified by Esposito. But auto-immunity is retained as a figure of necessity, not as a figure of all that goes wrong with immunity. Auto-immunity takes on a negative connotation in Esposito's work. For Derrida, it can have a positive connotation: it can, as we will see, refer to the way in which democracy (impossibly) can take place only by not taking place. It can also refer to unpredictable transformations of the reserves of inherited language – with results which may be negative or positive. For example, in *Rogues* Derrida describes as "auto-immune" the perversion of democracy which occurs when a culture and political language ("in line with a Greco-European ideal") are violently imposed by a colonizing nation on a colonized people. Colonialism might be promoted as a means of defending the interests of Europe and its democratic heritage and values. But, promoted by means of the imposition of a foreign legal system, administration, and educational system, by means of colonial violence, the acquisition of territory, the disappropriation of peoples and the failure to accord them the full rights of the colonizers, the democratic project is not only defended but also undermined by its colonialism. Colonialism is a hegemonic expression of democracy's expansionist ambitions which could be understood in this sense as its own auto-immunity: it undermines its own values as it seeks to extend its territory. But in the same discussion Derrida points out that the resulting colonial order can also be understood, in a more promising and optimistic sense, as auto-immune. For the measures which defend the colonizer's interests (the violent imposition of a Greco-European culture and political language) may also do double-duty in offering new conceptual and contextual resources for the countering political claims by colonized peoples to equal status, equal representation, or independence, sometimes leading to resistance, civil war, reform, or independence (*Rogues* 34–35).

II derrida's auto-immunity

Discussing his usage of the term auto-immunity in *Philosophy in a Time of Terror* (and referring to his "Faith and Knowledge"), Derrida describes his proposal "to extend to life *in general* the figure of an autoimmunity whose meaning or origin first seemed to be limited to so-called natural life or to life pure and simple, to what is believed to be the purely 'zoological,' 'biological' or 'genetic'" ("Autoimmunity" 187 n. 7).[2] How can we understand the divergence between Esposito's and Derrida's concept of auto-immunity such that Derrida can include an affirmative variant of the term, where Esposito will not?[3]

In one of Derrida's variants, cultural, psychic or security mechanisms are described as containing the *possibility* of becoming the vehicle for harm[4] to what they defend. This is closest to some understandings of the biological model of auto-immunity where, similarly, there is an occasional turn by a defensive system against the entity it otherwise protects. So we see Derrida's description of the West producing many of the devices (training techniques or weapons) used against it by terrorists, most obviously in 9/11 ("Autoimmunity" 115).[5] Similarly the availability of flight training schools, of an extensive network of domestic air travel and the predictable routines of airports would be considered components defending an American way of life which *also* provided the capacity for a means of attack on America by terrorists:

> it is perhaps because the United States has a culture and a system of law that are largely democratic that it was able to open itself up and expose its greatest vulnerability to immigrants, to, for example, pilots in training, experienced and suicidal "terrorists" who were [...] trained on the sovereign soil of the Unites States [...] The "terrorists" are sometimes American citizens [...] they received help in any case from American citizens; they took American airplanes, took over the controls, and took to the air in American planes and took off from American airports. (*Rogues* 40)

In what could be distinguished as a second model of auto-immunity, Derrida describes the intensification of attack provoked by the intensification of the defense mechanism – according to this variant of the immune mechanism, the stronger the defense mechanism the greater the consequent intensification of risk or danger. For example, in "Autoimmunity" Derrida describes the auto-immune model in terms of psychoanalytic understandings of repression and trauma ("all these efforts to attenuate or neutralize the effect of the traumatism (to deny, repress, or forget it, get over it) are but [...] so many autoimmune moments. Which produce, invent and feed the very monstrosity they claim to overcome" ("Autoimmunity" 99). He describes this stimulation of what one means to defend against (the "perverse effect" of "autoimmunity") as having both psychoanalytical and political registers:

> What will never let itself be forgotten is thus the perverse effect of the auto-immunitary itself. For we now know that repression in both its psychoanalytical senses and its political sense – whether it be through the police, the military, or the economy – ends up producing, reproducing and regenerating the very thing it seeks to disarm. (Ibid.)

Derrida also proposes a number of geopolitical tactical variants, such as smart bombs which provoke their victims to respond all the more strenuously, and terrorist attacks on America stimulated rather than deterred by the defense mechanisms against terrorism (ibid. 100). Esposito, describing a crescendo of immune mechanisms, has also argued that these follow the movement of acceleration. But he is describing the trends associated more specifically with late modernity as biologically oriented immune mechanisms have increasingly come to dominate political realities and imaginaries.[6] He is in agreement with Foucault, as Derrida is not,[7] with respect to the location of a threshold of new techniques and new political modes, a critical shift towards political formations concerned not only to maximize life but in extreme variants (and Esposito stresses the growing prevalence of these) to defend human life from contamination (*Terms* 130).[8]

According to a third variant of Derridean auto-immunity, where a defense mechanism aspires to immunize against a threat which appears exterior but is in fact constitutive, the terror against which steps are taken will always have been within: in this sense "the enemy is *also always* lodged on the inside of the system it violates and terrorizes" ("Autoimmunity" 188). This time Derrida refers to a defense mechanism which must inevitably turn, or "already will have" turned, on what it putatively defends. Here the point is not that a defense mechanism *could* turn self-destructive (or, as in the second variant, that it is likely to do so with ever increasingly velocity). Rather, on the third model the defensive mechanism by virtue of its very existence has already destroyed what it ought to have defended. On the first variant: this "could happen"; on the second variant, if it "could" happen, it will do so in the movement of crescendo and stimulation. On the third variant, the destruction in question has "already happened." It is not a defense mechanism that could harm, it is a defense mechanism which is *always already* the harm in question.

Examples in Derrida's work are seen, for example, in "Faith and Knowledge," when

Derrida describes an "autoimmune rejection of technoscience by means of technoscience" (Naas 203). Religion turns to technoscience to defend itself against the contaminations of technoscience. The techniques intended to offer immunity will already be examples of the harm against which defense is sought: thus auto-immune. Like the extensive appeal to technology to disseminate a message about the dangers of technology, further examples would be seen in the use of violent measures to protect against becoming a violent society. One of Derrida's best-known examples concerns the suspension of the Algerian elections in 1992, conceivable as a setting aside of democracy in the interests of defending democracy. To protect democracy against being undermined by its own process, democracy is set aside (the very danger against which the measure is intended to protect):

> The electoral process in Algeria would be [...] typical of all the assaults on democracy in the name of democracy [...] They decided in a sovereign fashion to suspend [...] democracy for its own good, so as to take care of it, so as to immunize it against a much worse and very likely assault. (*Rogues* 33)

If these are variants in which defense mechanisms are concurrently (and sometimes undecidably) defensive and harmful, they may be variants whereby harm and good, protection and defense, blur conceptually to the point of becoming indistinguishable from each other.[9]

Both Esposito and Derrida agree that immune mechanisms must fail, and all the more so the more vigorously they are enforced. Yet they account differently for this failure. Derrida's conceptual point of reference for this failure is the spacing, differing, and deferring of *différance*, seen when some kind of measure, some *pharmakon*, offers both "cure" (seeming to restitute a deferred presence) and "poison" (the presence it would restitute must always be deferred). Immune mechanisms will be those of the *pharmakon*, they will constitute both poison and cure. This gives one sense in which, for Derrida, immunity is really a form of auto-immunity. But while, and Esposito agrees, immunity must fail and must be identified as auto-immunity, we saw one sense in which Derrida *also* describes potentially positive senses of auto-immunity (the pervertibility of democracy *may* amount to the perfectibility of democracy – but the reverse is also true).

Much earlier, in *Positions*, Derrida had commented that while *différance* connotes the differing and deferring which suspends figures of putative originality or ideal identity or subjectivity (whether they pertain to nature, meaning or psychic life) it also is the condition for effects of subjectivity and meaning, although through their own paradoxical non-taking place, and their never being fully present.[10] In a similar sense, Derrida is also able to propose in *Rogues* that democracy be thought in terms of *différance*, which can be understood as auto-immunity. As Juliane Rebentisch presents this component of an argument from *Rogues*, since a thorough democracy is always postponed "democracy is securing itself by restricting itself, thereby threatening its own principles. This is why democracy is never really what it is" (Rebentisch 928).[11]

Again, auto-immunity is, in Derrida's work, given a potentially positive connotation, as when it can be said that democracy can only take place by never fully taking place: by restricting itself, by necessarily threatening itself. Hospitality and the pardon would also, in this sense, be their own auto-immunity, only able to take place by not entirely taking place, or by simultaneously counteracting themselves.[12] Thus, Derridean conditionality (elaborated in his discussions of hospitality, the gift and the pardon) would also offer another way of thinking about auto-immunity. Insofar as a limit is set to democracy, to tolerance, or indeed to hospitality in such examples, they restrict themselves, threatening their own principles. But auto-immunity could be said to allow them to "take place" (albeit in a paradoxical taking place occurring only through the work of differing and deferring, thus under the concurrent sign of their never fully taking place). Yet one should be wary of associating this idea with an overly positive reassurance: it is not that democracy decidably takes place *despite* never fully taking place. It is particularly challenging, in a deconstructive context, to do justice to the concurrent ideas of taking place "as/and" never fully taking place, but worth remembering Derrida's intermittent stress on the devastating ways in which democracy may not fully take place (his examples include the denial of rights to disenfranchised groups, the suspension of elections, political forms of racism, Eurocentrism, colonialism) despite its being true that it could not fully take place.[13]

III community, violence and the immune paradigm

In *Echographies*, Derrida explains his general reservations about the term community and its schemas of identity and unity (66). In *Of Grammatology*, he had deconstructed Rousseau's ideal of community: what Rousseau thinks of as community could only take place through the spacing, differing and deferring which appear to impede it. It seems as if large dispersed cities are an impediment to the small communities whose decision making could be accomplished by face-to-face negotiation. But the distance and alienation attributed by Rousseau to large dispersed cities pertains to all communities, including the smallest, no matter how physically proximate and intimate are the participants. Like democracy in its incompletion, community can be thought of as only possible through the *différance*, the delay, the deferred presence, in this case the lack of immediacy and spacing that seem to constitute its impediment. *Différance* would then be the auto-immunity of community in this "constitutive" sense.

For Derrida there is no account of community that could be otherwise described. We need spacing, deferring, differing, hollowing out to have anything approximating community. If he objects to the term "community" it is because it partakes of a language of the common, unity, and presence. If he refers to community, then, it will have to be a notion of the comm-auto-immunity, with concurrently negative and positive senses. So it is that in "Faith and Knowledge" (51) we find that he does refer to the term "auto-co-immunity" in a discussion of concepts of community which refer (undermining themselves as they do so) to the defense of blood, birth, nation, state, origin. This prompts the question of what kind of community could be affirmed: only a variant somehow affirmative of the necessity of its *différance*.

One of the points of convergence between Derrida and Esposito seems therefore to emerge insofar as Derrida understands community in these terms, and as only able to "take place" through not taking place. For Esposito also, although through a different means of theorizing its original lack and opening towards difference, community could only take place in its own absence from itself. It must be retheorized as the shared *lack* of the common and proper, with Esposito deriving analytic power from the image of having in common that we cannot have identity, self-presence or integrity in common, and that we are originally expropriated from ourselves. The starting position seems to be similar, with Esposito suggesting that anterior to any possible community is its *munus* – the opening to the other understandable in terms of gift, duty, obligation, the push towards externality, exposure, original expropriation. Certainly, as does Derridean *différance* (though by virtue of a different analysis, and different conceptual devices for theorizing alterity), Esposito's *munus* makes conceptually impossible a self-present or fully immunized community grounded in interiority and self-enclosure. A Derridean perspective concurs that we are best understood as united, not as in community but in our subtraction from an (im)possible community (and similarly, must be understood not as identities but as subtractions from (im)possible identities). If community takes place, they seem to agree, it is only in a paradoxical sense, only in and through such ongoing subtractions (paradoxical subtractions, of course, given that we are subtractions from no possible unity), and this would offer another sense in which one could say that community can only take place through not taking place, or, through not fully taking place.

Moreover, Esposito will also refer to Derridean *différance* in ways which make it sound closer to his own account of immunity:

> As Derrida has argued in a form that reinstates the logic and semantics of the immune lexicon, the *pharmakon* is what is opposed to its other not by excluding it but, on the contrary, by incorporating and vicariously substituting it [...] The *pharmakon* is both the evil and what opposes it [...] it is itself to the extent it is other [...] above all [the *pharmakon*] relates to life from the ground of its reverse. More than affirming life it negates its negation. (*Immunitas* 127)

But for Esposito an affirmative biopolitics, an affirmation of life, would offer an *alternative* to an immune logic here associated with the logic of the *pharmakon*, or *différance*. A theorizing of community in terms of *différance* alone would, from Esposito's perspective, lack the transformative conceptual resources that he attributes to an alternative, affirmative biopolitics. And while Derrida's rethinking community in terms of a necessary *différance* does seem equivalent as an analytic tool to Esposito's thinking community in terms of a necessary *munus* (openings towards the other, to be affirmed as such, rendering autonomy and purity impossible, violent aspirations), Esposito nonetheless associates Derridean *différance* and the *pharmakon* too specifically with the negative variants of a destructive immune paradigm, to also associate it with a (albeit entirely unpredictable) positive potential, as Derrida does.

So the effect of Esposito identifying *différance* with the movement of immunity is as follows. When he comments that

> as the largely fundamentalist and xenophobic currents of some neocommunitarian groups demonstrate today, nothing has a stronger immunizing effect than a community that completely "owns" itself, and which is then not common with respect to what does not belong to it (*Immunitas* 100),

we recognize that this is a remark akin to the form of Derrida's wariness about identitarian politics – for example, in *Monolinguism of the Other*, *The Other Heading* and elsewhere.[14] But whereas Esposito sees immunitarian violence manifesting in the vicious denial of *munus* and the denial of a shared *lack* of belonging, Derrida associates immunitarian violence with the disavowal of *différance*. Perhaps the terminological and conceptual disagreement on this point is best seen in Esposito's acknowledgements that a degree of the immune paradigm is conducive to life, but beyond a certain threshold it will only be increasingly destructive, thus auto-destructive and thanatopolitical (*Terms* 61). Notwithstanding Esposito's own discussion of the *pharmakon* in terms of "modern immunitary procedures" in this same passage from *Terms of the Political*, this is not the case for Derrida. In other words, while Derrida certainly describes *différance* as needed for life (just like the subject, and democracy, life can be defined as "divided from itself, in becoming space, in temporizing, in deferral" (*Positions* 29)), one would not say of *différance* (as Esposito says of immunity) that "it is needed for protecting our life, [but] if carried past a certain threshold, winds up negating life." There is no conceptual threshold *beyond* which *différance* converts into violent excess. Rather, it is the denial of *différance* (the aspiration to self-enclosure, self-identity) which amounts (and leads) to violent excess. For Derrida *différance* (in this sense, auto-immunity, a life which could only take place through its own not fully taking place) is essential to a life always mediated by death, and violent, fundamentalist, xenophobic, and "self-owning" aspirations to identity manifest in the disavowal of this necessity.

In sum, while both philosophers agree that immune paradigms lead to violent excesses in their attempts to defend and protect against risk and contamination, Esposito saves the term auto-immune for those violent excesses, the deadly or auto-negating extremes. For Derrida we saw it is a more general term, including some of what Esposito includes under the immune paradigm, but also the original *différance* perhaps most in proximity with (but not conceptually identical to) what Esposito might associate with *munus*. While it is true that *différance*, like *munus*, offers both philosophers viable ways of understanding the possibilities and realities of community, we see the divergence between them in the fact that for Derrida this never fully taking place is *best* understood as auto-immune, and in the fact that auto-immunity is understood by Derrida in *potentially* positive senses, and not just as the destructive excess and inclinations towards subordination, and defensive violence, of the immune paradigm. Because Derrida's stress is just as much on the negative variants of auto-immunity, the accounts of auto-immunity offered by Derrida and Esposito may intermittently seem close. But Esposito describes a violent collapse *from* immunity *to* (in *Esposito's* sense) the extreme of auto-immunity, whereas for Derrida immunizations will *always* have been (in *Derrida's* sense) auto-immune. Moreover, that fact is to be affirmed as having a transformative potential, and as having what Esposito identifies in Derrida's work as a messianic sense: amounting to an opening to an unknown, contentless future.

IV life, and the thanatopolitical extreme of the immune paradigm

I turn now to Esposito's accounts of the extreme of the immune paradigm, so as to begin situating the figures of pregnancy and birth in these contexts. We have seen that Esposito describes a model according to which "life cannot be preserved except by placing inside it what it is meant to counter or that which is meant to counter it" (*Immunitas* 8–9). The error of immunity would be seen in the attempt to *preserve* life – in the attempt to preserve stasis, to defend against transformation, deny the necessity of the homeopathic, and in the attempt to defend against what is figured as the threat from without. Describing the violent excesses associated with preservational effort, the term immune is used (though not only) to describe killing a part of oneself as a means of defending oneself, leading to the thanatopolitical extreme that one's own death might be associated with the defense of one's life. Esposito describes the specificity of the Nazi thanatopolitical extreme of biopolitics seen with the complete collapse of the division between biology and politics (*Bíos* 112). The furthest extremes of thanatopolitical defenses of life are also seen in the last days of the Nazi regime, with the promotion of the collective suicide of the entire German people *as* the means of preserving the German people (ibid. 111).[15] But Esposito's *Bíos* also opens with contemporary instances of (extreme) claims that life may sometimes be best defended by its own termination or forestalling: a case before the European Court in which a severely disabled child is represented by legal arguments that its life would best have been defended by the abortion his mother would have preferred; the murder by a Russian commando unit of hostages held by Chechen terrorists under the pretext this best defended their lives; and "humanitarian" bombing in Afghanistan, where the argument was put that lives were lost in the defense of the interests of those same lives.

Certainly, some of the examples recounted in the Introduction to *Bíos* could be otherwise depicted, but Esposito is particularly interested in contexts where the complete destruction of an entity is presented as the means of its putative defense. Not denying the many instances of biocracy in which racism's sub-divisions create caesurae within the "*biological continuum* addressed by biopower" (*Society Must Be Defended* 55) in the form of sub-groups directly or indirectly harmed as collateral damage, acceptable margins, or the object of explicit genocide in the supposed interests of the remainder, Esposito is particularly interested in a variant exemplifying the immune paradigm where "life" is described or describable as annihilated to preserve "life."

Among these, Esposito also considers *reproductive* variants of the violent and thanatopolitical excesses of biopolitics. Consider the last chapter of *Bíos*, as he turns to the reproductive thanatopolitics of German National Socialism – understood by Esposito in terms of the immunitary paradigm doing its worst:

> What the immunitary paradigm adds is the recognition of the homeopathic tonality that Nazi therapy assumes. The disease against which the Nazis fight to the death is none other than death itself. What they want to kill in the Jew and in all human types like them isn't life, but the presence in life of death: a life that is already dead because it is marked hereditarily by an original and irremediable deformation: the contagion of the German people by a part of life inhabited and oppressed by death. (*Bíos* 137)

Jews were deemed the internal enemy of Aryan aspirations but Esposito focuses on the Aryan nation understood as destroying a part of "itself"[16] – connecting this with a number of the examples of immunitary excess offered in the Introduction to *Bíos*, which share the ambiguous paradox of immunity. Similarly Esposito can present as a possible formulation that "Afghan lives" are killed as a means of defending "their" lives.[17] He chooses to focus on the possibility of killing a part of one's own state not as race bifurcation but instead as self-destruction. We see this particularly in the discussion of sterilization. Thus Esposito moves from this:

> Life is now defended by the state against every possible contamination. Racial hygiene is the immunitary therapy that aims at preventing or extirpating the pathological agents that jeopardize the biological quality of future generations (*Bíos* 128);

to the claim that sterilization is therefore

> the most radical modality of immunization because it intervenes at the root at the original point in which life is spread. It blocks life not in any moment of its development as its killer but in its own rising up – impeding its genesis, prohibiting life from giving life, devitalizing life in advance. (Ibid. 132)

Elaborating, Esposito comments that "it might seem paradoxical wanting to stop degeneration (whose final result was sterility) through sterilization." It seems paradoxical only if one fails to realize that such paradox (also presented by him as the negative doubling of the negative) is "an essential part, indeed the very basis of the immunitary logic itself" (ibid.).[18] Thus when the immune paradigm is associated with a suicidal tendency, one of the most extreme manifestations is identified by Esposito (in an account not to be found in either Foucault or Derrida) as the forestalling of life before it begins.[19] Forced deaths associated with eugenic programs are not only, in their thanatopolitical expression, the deaths of those killed but are also, as he puts it, a programmed annihilation in advance of the lives to which the murdered lives could have given life (ibid. 143–45).

Derrida does not elaborate a reference to life forestalled in advance. But when he does turn to nationalist and ethnocentric attempts to defend and protect purity, identity, origin, and also stresses their trend towards increasing violence, he includes the turn to eugenics in this context. For his part he describes efforts of self-immunization which he attributes to fundamentalist religions which perceive threat in modern technologies, as when

> the so-called return of religion tries to go back to the literality of idiom, the proximity of home, the nation, the earth, blood, filiation, and so on. In order to spirit away the threat, you therefore incorporate it in yourself, by appropriating technology, telecommunications, internet access, the effects of globalization, and so on. A process of self-immunization. It destroys the organism that it thereby seeks to protect. (*Paper Machine* 117)

The immunizing, ethnically or racially inflected desire to secure national boundaries in a number of ways will similarly manifest its violent, auto-immune form. In response to Le Pen's ideal immigration policy which could defend France against threats posed by "non-beneficial" immigration (immigrants considered to threaten homogeneity, social harmony or economic health), Derrida points out that such a filtration cannot be calculated in advance, and must be countered with an affirmation of unpredictability with respect to the *arrivant*. The filtering aim expresses a "fear of being altered by what comes from outside, by the other" (*Echographies* 19) – a fear which is not only destructive but auto-destructive. Derrida continues with a *countering* and *affirmative* stress on this auto-immunity:

> If it were capable of calculating this filtration in advance, a living organism might achieve immortality, but in order to do so, it would have to die in advance, to let itself die or kill itself in advance, for fear of being altered by what comes from outside, by the other, period. (Ibid. 18–19)

Derrida is quite willing to argue, as does Esposito, that the more extreme the attempts at self-immunization from what is perceived as change, transformation, invasion, the greater the auto-immunity (here in the sense of both destruction and auto-destruction). He continues, in the above passage from *Echographies*:

> Hence the theater of death to which racisms, biologisms, organisms, eugenics are so often given, and sometimes philosophies of life. (19)

At the point Esposito and Derrida mention the phenomenon of an auto-immune eugenics, the projects again seem to come into a kind of proximity, as they describe the theatre of death produced by such attempts at immunitary defense. Both remind us that these attempts will have included reproductive variants: violent or deadly forms of forced sterilization, genocidal rape, eugenic ambitions targeting women's reproductive lives, projects aimed at forestalling or forcibly determining "life" in advance) – as auto-immunity.[20]

V against big airs – politics and the organicist paradigm

Yet Derrida offers rejoinders which we will not see in the work of Esposito. Both agree about the dangers and violence of preservational and defensive logics. But, and characteristically, Derrida follows the above discussion with a further caution against a certain smugness that might attach to a widespread abhorrence on the left to Le Pen's propositions. For Derrida goes on to say that *anyone* who advocates border controls is complicit with this "organicist axiom," and, more strongly, that

> it is firmly rooted in the political insofar as it is bound up with and as long as it remains bound up with the nation state. (*Echographies* 19)

Derrida's proposal is that all who recoil from Le Pen must nonetheless refrain from putting on "big airs":

> let us underscore again what can't possibly make anyone happy: anyone, whether he is on the left or on the right, who, "like everybody," advocates immigration controls, bans illegal immigration, and would regulate the other, subscribes de facto and de jure, whether he likes it or not, and with varying degrees of elegance or distinction, to Le Pen's organicist axiom, an axiom which is none other than that of a national front. (The front is a skin, a selective "membrane", admitting only the homogenous or the homogenizable, the assimilable, or rather, that which is heterogeneous but considered to be "beneficial": the appropriable immigrant, the immigrant who is clean and proper.) We should not hide our eyes before this undeniable complicity [...] And when, like everybody, we have to acknowledge that we have no choice but to try to protect what we think is our inviolate body, when we want to regulate immigration and asylum, we should not put on airs and give lessons in politics [...] (Ibid.)

Esposito also recommends against the defensive mechanisms, the attempts to establish filtering membranes to achieve a rigorous defense of (an impossible) internal purity of community. He would also think we must be vigilant when it comes to the ongoing way in which membranes reassert themselves. Yet Esposito does not take pains, in the face of his critique of the immune paradigm, to argue so trenchantly that for all our analyses of its violence and thanatopolitics we cannot hope but to partake in membranes *in the potentially disastrous ways described by Derrida.* We can't entirely immunize ourselves from every way in which Le Pen would attempt to establish defensive membranes. For his part, Esposito argues for a reconfiguration of community. But the image offered by his work is of a vitalist and above all a *possible* reconfiguration. He does return us to an equivalent of Derrida's reminder that we must always look for, and acknowledge, our membranes, but Derrida's version affirms this necessary permeability *as* potentially terrible. This is not to be quietist or resigned to the monstrous possibilities arising from permeability, but it is to suggest that no political position could be immune from them:

> To acknowledge this permeability, this combinatory and its complicities, is not to take an apolitical position [...] On the contrary, it is to appeal to our duty to courageously formulate and thematize this terrible combinatory. (*Echographies* 19–20)

By contrast, Esposito undertakes to offer alternatives for this terrible combinatory – this is a critical point of differentiation between the figures of immunity and auto-immunity in their respective work. For Derrida one would, at best, commit to a politics of perpetual negotiation with a membrane constantly reinstalling itself – sometimes with terrible results, thus this is a recognition and negotiation said to call for "courage." Because for Esposito the most disastrous outcomes are the results of excessive attempts at immunization, an ideal for living with alterity is associated with alternatives that would not attempt radical immunization. Derrida has a rejoinder here, seemingly not shared by Esposito: the results of excessive immunization certainly are disastrous but we *cannot* not immunize. We cannot hope to secure ourselves from the disastrous results of immunization. The liberal cannot hope to secure him or herself from all of Le Pen's presuppositions.

Is it possible that Esposito, identifying the disastrous results of attempts at immunization (on which both agree), hopes, through a countering alternative model, to immunize against them? Derrida's perspective, by contrast, might prefer the emphasis that there can be no such immunization against the excesses of immunization. Yes, the immunizing membrane is a terrible combinatory. Yet no model, from Derrida or Esposito, could immunize against the potentially catastrophic consequences of attempts at self-protection.

One overlapping sense of auto-immunity in their work is the equivalence of total immunity with total annihilation. I suggested that both Esposito and Derrida agree that to be fully immunized would amount to death. Yet both remind us that no one lives without their defensive membranes.[21] So Esposito affirmatively rethinks the immune system as a "diaphragm" or "internal resonance chamber" through which difference "traverses us" (*Immunitas* 18). But Derrida is more inclined to remind us that the *worst* aspects of membranes will also reinstall themselves. A Derridean approach (to human rights claims, to feminism, to nationalism, to ethnocentrism, to racism, humanitarianism) will always provoke the question: where are our membranes? How do they operate? Where do they reassert themselves? At whose expense?

By contrast, Esposito turns to an alternative imaginary, in the sense that it is less infused with Derrida's vigilant wariness with respect to political pretensions to thoroughly transformative politics. *Immunitas* proposes a model of living with alterity in which *communitas* is not *inevitably* exposed to a future which might be terrible. Its exposure to alterity may well (in Derrida's sense) call for "courage," but Esposito does not affirm the inevitability of a potential for monstrosity. It is not, in *Esposito's* sense of the term, inherently nor inevitably auto-immune.

What, then, are the implications for the ways in which birth is represented in the work of the two philosophies of immunity?

VI birth

Every racism and ethnocentrism is, Derrida argues, regulated by means of defining the "alterity of the foreigner or the barbarian" through discourses on birth (*Politics* 91). In fact, a number of the examples discussed both by Esposito and Derrida (reproductively oriented ethnic rape, forced sterilization or abortion in the context of eugenics programs) also concern the deployment of women's capacity for birth to ethnocentric ends in biopolitical contexts. Thus to return to Derrida's comment in "Faith and Knowledge," when Derrida discusses the turn of ethnic-religious fundamentalisms to violence, he specifies:

> in the most lethal explosions of a violence that is inevitably ethnico-religious – [...] on all sides, women in particular are singled out as victims (not "only" of murders, but also of the rapes and mutilations that precede and accompany them). ("Faith and Knowledge" 49)

Both philosophers identify women, particularly women historically politicized and problematized as reproductive, as profoundly implicated in logics of racist and reproductively inflected immunity. Where Derrida does so directly and explicitly, Esposito's variant is less explicit but emerges in some of the examples discussed in *Bíos*.

And both philosophers turn to the figure of birth for alternative conceptual resources, but as they do so the differences I have discussed here again make themselves felt. In Esposito's case birth provides an alternative image of the mother–fetus relationship and of the new. In Derrida's case birth will provide images of hospitality and democracy to come.[22] Describing radical hospitality, Derrida offers as one figure of the *arrivant* the birth of a child:

> With the birth of a child – the first figure of the absolute arrivant – one can analyze the causalities, the genealogical, genetic, and symbolic premises, or the wedding preparations. But even if such an analysis could ever be completed, one can never reduce the element of chance [*aléa*], the place of the taking-place; there will be someone who speaks, someone irreplaceable, an absolute initiative, another origin of the world. (*Echographies* 20)

Another associated image offered concerns the immigrant, as seen in the reminder offered by Derrida to counter Le Pen:

> The immigration of which France's history is made, the history of its culture, religions and languages, was first a history of these children, children or immigrants or not, who were so many absolute *arrivants*. (Ibid.)

Though Derrida does not, in this reference to the child, discuss the danger posed to the host by the *arrivant* (as Royle notes, the *arrivant* is "hospitality itself" (Royle 111)), he does, in a number of his discussions of hospitality, emphasize the danger constitutively posed by the *arrivant*. In his discussions of unconditional hospitality, he stresses the possibility of one's home being destroyed ("Hospitality" 70–71). He depicts the future, associable with unconditional hospitality to the *arrivant*, as potentially monstrous (Derrida, *Points* 386–87). He establishes the interconnection: hospitality–*arrivant*–possible monstrosity, and separately, *arrivant*–child. As he does so he links together, via their respective association with *arrivance*, the child with the figure of monstrosity, if indirectly.[23]

If, a number of times in Derrida's work, the figures of birth and of the child allow him to evoke indirectly and directly the dilemmas of the impossible hospitality offered to the unknown *arrivant*, how does Esposito, for his part, draw on the fetus and the child, and also on literal and non-literal references to "birth"? How is his mode of doing so emblematic of the divergence between Derrida and Esposito? *Terms of the Political* continues to evoke birth as an emblem of Esposito's definitions of community as an opening towards otherness (44). Community, he writes, is "recalled in every birth" (ibid. 43), and similar references to birth are to be found in *Bíos* and *Immunitas*. But the latter work also includes a more specific reconfiguring of an imaginary for community associated with the mother–fetus relationship in *Immunitas*.[24] Esposito defines community as

> being traversed by a relation with others in such a way that we are altered: by the "with," the "between" and the threshold where [we] meet in a point of contact that brings [us] in contact with others to the degree to which [we] are separated from [our]selves. (*Communitas* 139)

He claims an emblematic importance for the reconfiguration of the mother–fetus relationship, one which could encompass his ideal reconfiguration of the immune paradigm:

> this is the ultimate – and prime – issue around which the entire immune paradigm wraps itself until reaching the point where it becomes indistinguishable from its opposite, "community": the force of the immune attack is precisely what keeps alive that which it normally should destroy. (*Immunitas* 171)

Rejecting the view of mother and fetus living in a harmonious relationship, or one of homogeneity, Esposito presents alternative images of the fetus and mother as always and necessarily at war with each other. He does not repudiate the view of the mother's body as attempting to reject the fetus, acknowledging that it can be said of the relationship that

> the mother is pitted against the child and the child against the *mother*. (Ibid.)[25]

But he emphasizes that the "spark of life" (ibid.) arises from this conflict. Reconfiguring the "war" and "defense" images of mother–fetus immunity, offering us an image of the fetus as (im)proper to the mother and vice versa, Esposito presents an image of the mother as able to tolerate the fetus, only *insofar* as it is other and heterogeneous. The image of interest to Esposito is the rejection of the fetus by the mother's body if it is genetically too *similar* to the mother. The genetic difference of paternal chromosomes stimulates an immune reaction which in turn stimulates a system of antibodies allowing the mother's body to retain the fetus (moreover, Esposito recalls that the presence of the fetus will in fact be therapeutic for women with autoimmune diseases (ibid.).[26]

This image offers a different means of conceptualizing immunity – it affirms the necessity or productivity of immune responses the absence of which would be catastrophic for life. They become the figure of a cohabitation with difference, an emblem for a different means of thinking about community. This is most literally introduced with a reference to biologists who think of the immune system as an ecosystem or a social community – not just a defensive system, but a system of "self-alteration" (ibid. 169; see also 177).[27]

So, how do the respective references to birth by Esposito and Derrida help us to differentiate their account of immunity?

VII birth and destruction

Derrida associates the figure of the *arrivant* with birth, and with the child. But he also associates the arrivant with the figure of a radical hospitality, an unpredictable future, and with the possibility of the monstrous. In such contexts, his reminder is that the *arrivant* can potentially do harm, be terrible, or "destroy your house" ("Hospitality" 70–71). Given these associations, the child (the figure, for Derrida, of the *arrivant*) ought to be, and almost is (with the caveat I have mentioned – for the connection is made indirectly), represented as potentially negative, destructive, revolutionary, deadly, monstrous, in unforeseeable ways.

Compare this to Esposito's image of the mother–fetus relationship. As a reconfigured image, it excludes the vagaries of pregnancy potentially destructive to the mother. He gives his attention to the fetal enhancement of the mother's immune system but not to phenomena threatening the mother's life. Accordingly, the phenomena doing service as emblematic figures of community and openness towards the other here will not include gestational diabetes, ectopic pregnancy, preeclampsia, and eclampsia – nor scenarios ranging from postpartum depression to matricide, or other negative inflections potentially associable with the unpredictable aspects of the fetus and pregnancy. This is not to favor a negative representation of maternity, but to indicate what drops out from both Esposito and (just slightly) even Derrida's version. But in Derrida's case, this is a missing connection in a series of associations already in his work. He does not emphasize the child's capacity to be monstrous for its host, but he does associate the *arrivant* with the child, and, separately, the *arrivant* with the possibility of monstrosity and destruction of the host. By contrast, Esposito places no particular emphasis on a monstrous *arrivant*. Rather, he favors an affirmative and vitalist account of the constant transformation of an entity's own norm.

This is seen not only in Esposito's account of life in *Bíos* but also in his account in that work of the termination of reproductive life (in the form of eugenically oriented sterilization or murder) as an extreme thanatopolitics, cutting off life in advance. Derrida can answer Esposito here. For Derrida, such images will not serve. Dating back even to *Positions*, on Derrida's account life *must* be understood as having the potential to cut itself off in advance: to be its own thanatopolitics.[28] A number of commentators, including Lee Edelman and W.J.T. Mitchell, have criticized Derrida for a seemingly strange inconsistency in some of his work on auto-immunity. Edelman has argued that a futurism, and possibly a reproductive futurism (to use Edelman's term), reinstalls itself fleetingly in his work ("Against Survival" 161–62). But Derrida's emphasis on monstrosity also offers the resources to counteract that tendency. This means he also offers the resources to challenge Esposito's images (without representation of exceptions) of children as provoking love even in the most thanatopolitical of contexts, as in the genocidal rapes of Rwanda (those mothers figured by Esposito as loving their children, despite the hateful circumstances of their conception (*Bíos* 7)) or in their heterogeneity as also able to promote the health of the mother ("the child's attack can also save the mother from her self-injurious tendencies" (*Immunitas* 171)).

Surely these are, as Derrida has said of the figures of birth, fraternity, and the association of nationality and maternity discussed in *Politics of Friendship*, both metaphors *and* non-metaphors:

> And any time the literality of these implications has been denied, for example by claiming that one was speaking not of the natural and biological family (as if the family was ever purely natural and biological) or that the figure of the brother was merely a symbolic and spiritual figure, it was never explained why one wished to hold on to and privilege this figure rather than that of the sister, the female cousin, the daughter, the wife or the stranger, or the figure of anyone or whoever. (*Politics* 58)

To whatever extent Esposito's images of the mother–fetus favor life (albeit as alternatives intended as a means of offering alternatives to the thanatopolitical variants), they can be similarly challenged. Why present only the image of the good done to the mother by the alterity of the fetus, rather than the concurrent scenario of unpredictable harm which also is possible? If we are offered an image of transformative or even therapeutic self-transformation, through images hovering between the literal and non-literal, why not include more diverse phenomena and diverse outcomes? In particular, what of Derrida's monstrous possibilities?

In sum, Derrida and Esposito agree on the hazards of immunitary defense and of preservational strategies, but offer different responses to the disastrous aspects of immune logics. Esposito turns to an ideal image of community, for which a reconfigured image of the mother–fetus is presented as providing a model, as is the diaphragm allowing difference to traverse us. When he cites an immunological account of the furious attack of the mother's system on the fetus, he does so in reference to its paradoxically allowing, rather than impeding, the fetus's survival, commenting:

> This is the ultimate – and prime – issue around which the entire immune paradigm wraps itself until reaching the point where it becomes indistinguishable from its opposite, "community": the force of the immune attack is precisely what keeps alive that which it should normally destroy. The mother is pitted against the child and the child against the mother, and yet [...] contrary to the metaphor of a fight to the death, what takes place in the mother's womb is a fight "to life", proving that difference and conflict are not necessarily destructive [...] From this perspective, nothing remains of the incompatibility between self and other. The other is the form the self takes where inside intersects with outside,

> the proper with the common, immunity with community. (*Immunitas* 171)

Exactly where Esposito aspires to a reconfigured immunity allowing an image of community as not necessarily destructive, Derrida will have been at his most cautious. An openness to the exterior is also an openness to destruction. This must be affirmed, not denied, for the very reason both philosophers acknowledge. There can be no life with "full immunity." While this is the condition of life, it also means that defensive membranes *in all their violence*, do re-establish themselves.

We have seen Derrida's suggestion that this needs to be thematized "with courage." My point is not that Esposito rejects this view. He is not far from Derrida in affirming a vigilant attention to the constant re-establishment of defensive membranes. But he looks for alternatives, for the possibility of an affirmative biopolitics, and this in lieu of affirming the necessary possibility of monstrosity and catastrophe at work in exposure to transformation, auto-destruction, the envelope, differing and deferring, self-limiting, the *à venir*, the ultra-transcendental, the not-taking-place, self-cancellation, and, most broadly, alterity. For Esposito, Derrida's remark that

> no community is possible that would not cultivate its own auto-immunity ("Faith" 51)

could only be a negative identification of a community pushed to the extremes of the immune paradigm. But for Derrida this also has a positive connotation: it is the matter of concurrent affirmation. We cannot live except enfolding exteriority. It is because we are open to disaster that we are open to transformation. But the reverse also must hold. If we are open to transformation, we are open to disaster.

I have argued that this different inflection in Derrida and Esposito manifests in a component of the latter's material little engaged by his commentators,[29] the limited, overly positive and overly "decidable" figures of mothers, pregnancy, and birth in Esposito's work. I have suggested that from a Derridean perspective, Esposito's related, affirmative defense of (reconfigured) community could be seen as a political and philosophical project to *immunize against* the dangers of immunization and auto-immunity. This characterization would similarly be suggested by the comparison with the Derridean model, which would maximally caution against the hope of securing positive alternatives to the dangers of immunization and the unpredictable work of auto-immunity.

notes

My warmest thanks to Greg Bird, Jonathan Short, and an anonymous reviewer for *Angelaki* for their critical and generous suggestions on earlier versions of this paper.

1 See, for example, Timothy Campbell's introduction to his translation of Esposito's *Bíos: Biopolitics and Philosophy* ("Bíos, Immunity, Life" vii–xlii); Vanessa Lemm's introduction to her translation of Esposito's *Terms of the Political* ("Biopolitics and Community in Roberto Esposito" 1–13); Amendola 102–18; Haddad 173–93; Wolfe 92ff.

2 In "Faith and Knowledge" Derrida refers to a biological auto-immunization that would seem to give rise to a more general logic:

> It is especially in the domain of biology that the lexical resources of immunity have developed their authority. The immunitary reaction protects the "indemn-ity" of the body proper in producing antibodies against foreign antigens. As for the process of auto-immunization, which interests us particularly here, it consists for a living organism, as is well known and in short, of protecting itself against its self-protection by destroying its own immune system. As the phenomenon of these antibodies is extended to a broader zone of pathology and as one resorts increasingly to the positive virtues of immune-depressants destined to limit the mechanisms of rejection and to facilitate tolerance of certain organ transplants, we feel ourselves authorized to speak of a sort of general logic of autoimmunization. ("Faith" 73 n. 27)

But Derrida considers auto-immunity not to arise from a biological context that, by analogy, could then be applied to the political sphere. Rather, "if autoimmunity is physiological, biological, or zoological, *it precedes or anticipates all these oppositions*" (*Rogues* 109; my emphasis). Derrida does not take a biological model and apply it more broadly (e.g., to religion, democracy and ethnocentrism); rather, he develops a way of thinking immunity and argues that it would precede and be presupposed by both a biological and political declension. One can similarly place in context the reference in *Terms of the Political*, where Esposito refers to the biological phenomenon of auto-immunity, likening the way in which the Nazi regime progressively widened its circle of death, "as in so-called auto-immune diseases, where the immune system is strengthened to the point of causing the body's decomposition" (73). For Esposito has also emphasized that "the biological immune is [not] a neutral or original object compared to the derivative or metaphoric nature of other forms of social immunization" (*Immunitas* 16).

3 In what follows, I discuss several versions of the term "auto-immunity" in Derrida's work. In *Immunitas* Esposito refers to Derrida's account of auto-immunity, but only to the version discussed in "Faith and Knowledge" (*Immunitas* 52–56)

4 This is discussed in contexts as varied as the autobiographical impulse, discussed in "The Animal That Therefore I Am," in which immunization through the autobiographical gesture *risks* becoming autoimmune. Commenting that "nothing risks becoming more poisonous than an autobiography" (415), Derrida describes "this terrible (*and always possible*) perversion by means of which the immune becomes auto-immunizing," an "immunizing movement that is always *threatened* with becoming auto-immunizing" ("The Animal" 415; emphasis mine), as well as the defense of America's "security" (see below).

5 See also Esposito's discussion of 9/11 in terms of an immunitary crisis (*Terms* 132).

6 See, for example, the account of this as culminating in a global crisis "at the culmination of the biopolitical epoch" (*Terms* 132–33).

7 See Derrida, *Beast* 324.

8 Esposito specifically sees Derrida as "not completely register[ing] or [...] reject[ing] the paradigmatic shift toward *bíos* that for some time now has altered the dynamics of knowledge" – this would apply both to the paradigm in which politics is increasingly concerned with the defense of an impossibly immunized life, and the growing epistemological and philosophical importance of theoretical accounts adequate to this phenomenon ("Contemporary French" 112–13).

9 This variant of Derridean auto-immunity emphasizes more strongly the constitutive, as when, in *Specters of Marx*, Derrida describes the living ego as auto-immune. It must, he argues, necessarily "welcome the other within." There is no living ego without: "iterability, non-uniqueness, prosthesis, synthetic image, simulacrum," all, he argues, "so many figures of death" (*Specters of Marx* 141). There could not be an ego without the alterity that might seem to be the ego's negation and other. Elsewhere he argues that there could not be life without the immanence of death, and the degradation that appears to be its antithesis.

10 In the 1968 interview with Julia Kristeva in *Positions*, Derrida describes "the conscious and speaking subject" as "not present, nor above all present to itself before *différance*." But, he adds, "the subject is constituted" (but, of course, he continues: "only in being divided from itself, in becoming space, in temporizing, in deferral" (*Positions* 29)).

11 Rebentisch associates this account with the conditionality of the liberal tolerance associated with some discourses of multiculturalism. Insofar as we are tolerant of what we tacitly understand to be (sufficiently) like us (insofar as our tolerance is tacitly premised on a degree of likeness), tolerance could be said (according to Rebentisch's reading) to similarly not take place, not to be what it is, to secure itself by restricting itself.

12 See Derrida, *Given Time* and *The Gift of Death*.

13 The latter does not, of course, justify the former, and it would be incorrect, also, to settle these coinciding formulae into a decidable version of democracy either "taking place" or "not taking place." Similarly, just as it is incorrect to suppose Derrida's auto-immunity should be understood as decidably destructive, it is incorrect to assume a decidable distinguishability between its destructive and positive possibilities or effects.

14 See, for example, *Monolinguism* 28, and his comments on cultural identity in *The Other Heading* and also in Caputo, *Deconstruction in a Nutshell*:

> take the example of a person or of a culture. We often insist nowadays on cultural identity – for instance, national identity, linguistic identity, and so on. Sometimes the struggles under the banner of cultural identity, national identity, linguistic identity are noble fights. But at the same time the people who fight for their identity must pay attention to the fact that identity is not the self-identity of a thing, this glass, for instance, this microphone, but implies a difference within identity. That is, the identity of a culture is a way of being different from itself; a culture is different from itself; the person is different from itself. Once you take into account this inner and other difference, then you pay attention to the other and you understand that fighting for your own identity is not exclusive of another identity, is open to another identity. And this prevents totalitarianism, nationalism, egocentrism, and so on. That is what I tried to demonstrate in the book called *The Other Heading*: in the case of culture, person, nation, language, identity is a self-differentiating identity, an identity different from itself, having an opening or a gap within itself. (*Deconstruction in a Nutshell* 13–14)

This remark must be counter-balanced, however, by comments made about Le Pen and the membrane; see *Echographies* 19, discussed in section V below.

15 This takes further Foucault's discussion in *Society Must Be Defended* of Hitler's April 1945 order "to destroy the German people's living conditions" (260), but see nn. 16 and 17.

16 There is room for debate about how closely this follows Foucault's account in *Society Must Be Defended* of the caesura introduced by biopolitics into the biological continuum. The break described by Foucault emphasizes the emergence of biological inferiority within a population in the form of divisions between races, sub-species, or "inferior" groups. He describes the emergence of rationales for letting these groups die, or be killed, to the ends of the healthy life of the population. Unlike wars between peoples and nations, he stresses that this is a conceptualization of internal elements *within* a population considered destructive or threatening to the future. Still, he emphasizes: "it is a way of separating out the groups that exist within a population." According to his commentary: "the death of the other, the death of the bad race, of the inferior race (or the degenerate, or the abnormal) is something that will make life in general healthier: healthier and purer" (255). Esposito, in his reference to the parallels made by Nazi biologists between sterilization, genocide and removing the appendix of the population, inclines his analysis towards the extreme version of the immune paradigm: where self-destruction eventually becomes a means of enhancing "life."

17 It would be just as plausible to argue that some Afghan lives are considered to be collateral damage to the overall "humanitarian ends" of other Afghans or other interests. In other words, Esposito's interpretation can be distinguished, in this emphasis, from that offered by Butler in *Frames of War* and *Precarious Life*, and, I have suggested, from that offered by Foucault in *Society Must Be Defended*.

18 Esposito considers this the extreme logic of the immune paradigm. The element of seeming paradox (though he does not favor this term) is emphasized insofar as he describes the attempt to avoid race "sterility" through mechanisms of "sterilization." The risk of sterility to the Aryan race was attributed to individuals associated with degeneracy or transmittable racial or biological inferiority. But Esposito cites Nazi medical analogies between Jews and the gangrenous appendix of the German *Völkerkörper* and the concepts of racial hygiene and demographic politics of Verschuer who defended a "healthy patrimony [through] the elimination of *its* sick elements" (*Bíos* 143).

19 See Anne O'Byrne's "*Communitas* and the Problem of Women" (this issue), for her discussion of Esposito on the "suppression of birth."

20 At the radical extreme, for Derrida, immunity from the risk of unpredictable transformation could only be accomplished by the most radical stasis: death. Thus any of the attempts to stave off foreignness and difference are bound to fail. They are, therefore, auto-immune, here in the sense of being destructive to their own aspirations.

21 For example, Esposito does acknowledge that "immunity is needed for protecting our life," provided it does not exceed the threshold beyond which it "winds up negating life" (*Terms* 61).

22 See Haddad's *Derrida and the Inheritance of Democracy* for its definitive discussion of the status, and multiple meanings, of birth in Derrida's work.

23 He is comfortable with the direct associations of *arrivance*, future, hospitality and monstrosity, but with some seeming preference for the more oblique association of the child with the monstrous. I add here to the discussion of birth in Derrida's work opened by Haddad, who comments:

> On the one hand, [Derrida] identifies [birth] narrowly with nature and necessity, and this identification grounds his resistance to the use of fraternity in political discourse. Birth understood in this sense is thus marked as negative, since it is said to be at work in a discourse whose force Derrida wishes to diminish. On the other hand, Derrida himself has recourse to birth as a figure for the *arrivant*, as that which comes and exceeds all calculation and determination, precisely in political contexts. Here Derrida imparts a positive value to birth, insofar as he invokes the figure of the child or its substitutes, suggesting he endorses an openness to the arrivant. Derrida's work would thus appear to contain two births, with two meanings and two values. (Haddad, *Derrida* 122–23)

24 Esposito hopes to add an alternative understanding of community, one offered to counteract the excesses of the immune paradigm and their thanatopolitics. By contrast, in a favored modeling of community we would not attempt to thoroughly immunize ourselves (or community) from risk, threat, death, difference, encroachment, transformation. Such attempts only aggravate, in any case, what they attempt to avert. We would not attempt to protect ourselves from this "no-thing" that is the mode of being of community (*Communitas* 139).

25 And similarly, see Esposito's interview with Timothy Campbell and Federico Luisetti, "On Contemporary French and Italian Political Philosophy," in which Esposito again uses an image of the "the internal struggle, which for example, compares and contrasts the fetus's immunitarian system with that of the mother" as an alternative for a biopolitics thought of as a "predetermined thread linking birth to death" (111).

26 Making this point, he draws particularly on a discussion of the fetus and the immune system presented by John Dwyer in *The Body at War*. Anne O'Byrne's contribution to this special issue offers further reflections on Esposito's interpretation of the paternal chromosomes, and of the fetus as a "semi-allogenic implant in the maternal body, preserved there by paternal difference" (O'Byrne this issue).

27 Esposito refers also to Donna Haraway's account of a semi-permeable self, enfolding within its boundaries the alterity of its environment (*Immunitas* 165–66).

28 As late as his *Death Penalty* seminar, Derrida exchanges an articulation of the "right to life" for an account of life *as* the undecidability of one's own death (*Séminaire*; see in particular the tenth session, 331–64).

29 But, for important exceptions, see Haddad, "Citizenship" and (for his treatment of Derrida on birth) idem, *Derrida*; see also O'Byrne's contribution in this special issue of *Angelaki*. In an earlier work, O'Byrne offers an analysis of the figure of birth by a number of contemporary and twentieth-century Continental philosophers, including Derrida; see her *Natality and Finitude*.

bibliography

Amendola, Adalgiso. "The Law of the Living: Material for Hypothesizing the Biojuridical." *Law, Culture and the Humanities* 8.1 (2012): 102–18. Print.

Butler, Judith. *Frames of War: When is Life Grievable?* London and New York: Verso, 2009. Print.

Butler, Judith. *Precarious Life: The Powers of Mourning and Violence*. London and New York: Verso, 2004. Print.

Campbell, Timothy. "Translator's Introduction: *Bíos*, Immunity, Life: The Thought of Roberto Esposito." *Bíos: Biopolitics and Philosophy*. By Roberto Esposito. Trans. Timothy Campbell. Minneapolis: U of Minnesota P, 2008. vii–xlii. Print.

Caputo, John D. *Deconstruction in a Nutshell: A Conversation with Jacques Derrida, Edited, with a Commentary by John D. Caputo*. New York: Fordham UP, 1997. Print.

Derrida, Jacques. "The Animal That Therefore I Am (More to Follow)." Trans. David Wills. *Critical Inquiry* 28 (2002): 369–418. Print.

Derrida, Jacques. *The Beast and the Sovereign.* Vol. 1. Ed. Michel Lisse, Marie-Louise Mallet, and Ginette Michaud. Trans. Geoffrey Bennington. Chicago and London: U of Chicago P, 2009. Print.

Derrida, Jacques. "Faith and Knowledge." Trans. Sam Weber. *Religion.* Ed. Jacques Derrida and Gianni Vattimo. Stanford: Stanford UP, 1998. 1–78. Print.

Derrida, Jacques. *The Gift of Death.* Trans. David Wills. Chicago: U of Chicago P, 1992. Print.

Derrida, Jacques. *Given Time: I. Counterfeit Money.* Trans. Peggy Kamuf. Chicago: U of Chicago P, 1991. Print.

Derrida, Jacques. "Hospitality, Justice and Responsibility." *Questioning Ethics: Contemporary Debates in Philosophy.* Ed. Richard Kearney and Mark Dooley. London: Routledge, 1999. 65–83. Print.

Derrida, Jacques. *Monolinguism of the Other or The Prosthesis of Origin.* Trans. Patrick Mensah. Stanford: Stanford UP, 1998. Print.

Derrida, Jacques. *Of Grammatology.* Corrected ed. Trans. Gayatri Chakravorty Spivak. Baltimore and London: Johns Hopkins UP, 1997. Print.

Derrida, Jacques. *The Other Heading: Reflection on Today's Europe.* Trans. Pascale-Anne Brault and Michael B. Naas. Bloomington and Indianapolis: Indiana UP, 1992. Print.

Derrida, Jacques. *Paper Machine.* Stanford: Stanford UP, 2005. Print.

Derrida, Jacques. *Points ... Interviews 1974–1994.* Ed. Elisabeth Weber. Stanford: Stanford UP, 1995. Print.

Derrida, Jacques. *The Politics of Friendship.* Trans. George Collins. London: Verso, 1997. Print.

Derrida, Jacques. *Positions.* Trans. Alan Bass. Chicago: U of Chicago P, 1981. Print.

Derrida, Jacques. *Rogues: Two Essays on Reason.* Trans. Michael Naas and Pascale-Anne Brault. Stanford: Stanford UP, 2005. Print.

Derrida, Jacques. *Séminaire. La Peine de mort. Volume I (1999–2000). Edition établie par Geoffrey Bennington, Marc Crépon et Thomas Dutoit.* Paris: Galilée, 2012. Print.

Derrida, Jacques. *Specters of Marx: The State of the Debt, the Work of Mourning, and the New International.* Trans. Peggy Kamuf. New York: Routledge, 1994. Print.

Derrida, Jacques, and Giovanna Borradori. "Autoimmunity: Real and Symbolic Suicides. A Dialogue with Jacques Derrida." *Philosophy in a Time of Terror: Dialogues with Jürgen Habermas and Jacques Derrida.* Ed. Giovanna Borradori. Chicago and London: U of Chicago P, 2003. 85–136. Print.

Derrida, Jacques, and Ann Dufourmantelle. *Of Hospitality.* Trans. Rachel Bowlby. Stanford, Stanford UP, 2000. Print.

Derrida, Jacques, and Bernard Stiegler. *Echographies of Television.* Trans. Jennifer Bajorek. Cambridge: Polity, 2002. Print.

Dwyer, John M. *The Body at War: The Miracle of the Immune System.* New York: New American Library, 1988. Print.

Edelman, Lee. "Against Survival: Queerness in a Time that's Out of Joint." *Shakespeare Quarterly* 62.2 (2011): 148–69. Print.

Edelman, Lee. *No Future: Queer Theory and the Death Drive.* Durham, NC: Duke UP, 2004. Print.

Esposito, Roberto. *Bíos: Biopolitics and Philosophy.* Trans. Timothy Campbell. Minneapolis: U of Minnesota P, 2008. Print.

Esposito, Roberto. *Communitas: The Origin and Destiny of Community.* Trans. Timothy Campbell. Stanford: Stanford UP, 2010. Print.

Esposito, Roberto. *Immunitas: The Protection and Negation of Life.* Trans. Zakiya Hanafi. Cambridge: Polity, 2011. Print.

Esposito, Roberto. *Terms of the Political: Community, Immunity, Biopolitics.* Trans. Rhiannon Noel Welch. Intro. Vanessa Lemm. New York: Fordham UP, 2012. Print.

Esposito, Roberto, with Timothy Campbell and Federico Luisetti. "On Contemporary French and Italian Political Philosophy: An Interview with Roberto Esposito." *Minnesota Review* 75 (2010): 109–18. Print.

Foucault, Michel. *Society Must Be Defended: Lectures at the Collège de France, 1975–76.* Ed. Mauro

Bertani and Alessandro Fontana. General editors François Ewald and Alessandro Fontana. London: Picador, 2003. Print.

Haddad, Samir. "Citizenship and the Ambivalence of Birth." *Derrida Today* 4 (2011): 173–93. Print.

Haddad, Samir. *Derrida and the Inheritance of Democracy*. Bloomington: Indiana UP, 2013. Print.

Hägglund, Martin. *Radical Atheism: Derrida and the Time of Life*. Stanford: Stanford UP, 2008. Print.

Haraway, Donna. "Biopolitics and Postmodern Bodies." *Simians, Cyborgs and Women: The Reinvention of Nature*. New York: Routledge, 1991. 203–30. Print.

Mitchell, W.J.T. "Picturing Terror: Derrida's Autoimmunity." *Critical Inquiry* 22 (2007): 277–90. Print.

Naas, Michael. *Miracle and Machine: Jacques Derrida and the Two Sources of Religion, Science, and the Media*. New York: Fordham UP, 2012. Print.

O'Byrne, Anne. "*Communitas* and the Problem of Women." *Angelaki* 18.3 (2013): 125–38. Print.

O'Byrne, Anne. *Natality and Finitude*. Bloomington and Indianapolis: Indiana UP, 2010. Print.

Rebentisch, Juliane. "Derrida, Democracy, and America." *Cardozo Law Review* 27 (2005): 927–33. Print.

Royle, Nicholas. *Jacques Derrida*. London and New York: Routledge, 2003. Print.

Wolfe, Cary. *Before the Law: Humans and other Animals in a Biopolitical Frame*. Chicago: U of Chicago P, 2012. Print.

Therefore, rather than referring to a body, an essence, or a common subject, community refers precisely to this game of relation and distinction, to the proximity of the interval and of proximity. This is the reason that, despite an account of the failure of all communisms and the danger of all communitarianisms, the community remains our question. One might even come to the point of posing our own unique questions, that which makes the escape from sense still our sense, the sense of "we."

Roberto Esposito in dialogue with Jean-Luc Nancy in "Dialogue on the Philosophy to Come" 80

ignaas devisch

HOW (NOT) TO PROPERLY ABANDON THE IMPROPER?

I the improper

Today, the improper is not only a philosophical issue; it is also a political question. Our era regularly witnesses political struggle with the improper. Consider, for example, the success of newly founded political parties in several European countries – Finland, Greece, Austria, Belgium, the Netherlands, to mention only a few. These parties make it part of the agenda to pursue authentic or proper being based on common norms and values whilst moving away from the improper elements of migration and cultural shifts. Thus, the question of the improper is clearly a contemporary issue. In particular, striving to abandon the improper is central to the contemporary political agenda in many Western countries. Often, the claim behind this appeal is that we previously shared a certain destiny or communality and that this transparent presence subsequently became lost in our time of pure appearance, externality, absence, and dislocation. Hence, the call for the possible recovery of this original and communal properness, and, furthermore, the idea of abandoning the improper, are both conceived as the solution to social problems.

At first glance, abandonment appears to be a straightforward and transparent idea: to enter a proper existence, simply leave the improper behind you. And, in fact, this idea of a "proper abandonment of the improper" is quite present in the outlines of several extreme right political parties in Europe and elsewhere: expel the foreigners from your country and the problems disappear along with them. Consider my own country, for example; in the 1990s an extreme right party in Flanders became very popular with the slogan "*eigen volk eerst*" (our own people come first). The idea that it might be possible to separate the improper (otherness) from the proper (ownness) became central to this party, which at a certain time had nearly 20 percent of the voters behind them (and this idea).

This idea or, as I will argue, illusion of not being related to some "improper outside" is central to this paper. As a category, however, the improper does suggest that one could possibly abandon it: after all, the improper is improper, and this is far from a mere linguistic pun. The question of the improper and its negativity is and has been of crucial importance for philosophy, particularly for a great number of Continental philosophers' thinking about community, being together, or related topics.

Given the risks of a political agenda abandoning the improper in a proper way and realizing a "closed community," contemporary philosophers such as Roberto Esposito and Jean-Luc Nancy, among many others, have thought about the relationship between the proper and improper. They not only discuss, dismantle or deconstruct the idea of a strong opposition between proper and improper; they also emphasize the improper within everything proper. Every proper existence is always marked by impropriety, Nancy and Esposito argue; there is no "original" or "proper" commonality that would liberate us from all things improper. Even before one can appropriate one's property, it is already traversed by the improper.

This argument is important to the analysis of the improper in this paper. The question "how not to abandon properly the improper?" is in fact one of the imperatives behind the arguments of Esposito and Nancy: if indeed the proper is always already, in one way or another, related to or contaminated by the improper, it is indeed an illusion to think we can simply leave the improper behind.

Both Esposito and Nancy argue that no single community can say of itself "I am who I am" or "I am my own properness." To explain this in terms that Jacques Derrida has made familiar, all originality or every proprium has always already escaped us, not least at the very moment we are looking for it or attempting to return to it. The "return to" creates the originary supplement in which we are always already stuck. Translated into the terms of this paper: one can never abandon the improper in a proper way. I will come back to this.

It is remarkable that Martin Heidegger is an exemplary figure in the analyses of both philosophers. The reason why this is quite remarkable is that Heidegger is not only known for his existential analysis but also for his political engagement in a destructive ideology which consisted of a proper community, one of its nodal objectives. In particular, if we think about the early writings of Heidegger, the idea of "authentic Being-with" or his ambivalence in *Being and Time* towards "the They" – where he believes he is able to separate the inauthenticity of "the They" from the heroic authenticity of a people – Heidegger seems to be a counterpoint to what Esposito and Nancy argue. Nevertheless, Esposito and Nancy put their fingers on Heidegger's sore spot. Their analysis of Heidegger's problematic thinking is, at the same time, a renewed effort to deal with the question of the improper. As a result, both philosophers develop their own thinking of community and identity, and we will present the key concepts of their philosophical framework. Furthermore, one can also argue with respect to Heidegger's writings that "*Jemeinigkeit*" (mineness) can never be understood as an individualistic self or a self-present individual (Raffoul, "Otherness" and "Logic"). Heidegger's critique of a certain form of Being-with, namely, the absorption of individuals in "the They," derives from his reservations about an identification-based being-together. Heidegger therefore leaves us with an ambivalent stance towards the proper and the improper, and in this paper I consider this ambivalence crucial not only to understand Heidegger but also to realize what is at stake today in our thinking the improper.

Since the reasons for Heidegger's political motivations have been discussed thoroughly (Farias; Lacoue-Labarthe; Janicaud, *L'Ombre*; Safranski; Polt), I will not revisit the issue here. Rather, I want to focus on the discussion of Heidegger's earlier work on Being-with and properness as articulated in *Being and Time*. This is also the main focus of Esposito and Nancy. I will analyze their discussion of Heidegger primarily with regard to four categories: on the one hand, the proper and improper (and their intertwining), and on the other, community and immunity. Not only are these four categories vital for illuminating what is at stake in Heidegger's gesture but also for understanding why both Esposito and Nancy, two crucial voices in this context, state that it is "impossible to abandon the improper in a proper way" in their writings on community (Campbell; Farneti; Huang).

Esposito demonstrates the original linkage between community and the founding risk of expropriation. From his perspective, community is anything but a common essence or a shared property. Community, for Esposito, inaugurates the very absence or expropriation of the proper, and the impossibility of the in-common as founded on mutual appropriation of the proper. This line of argumentation is very similar to what Nancy states in several of his books and articles: being-in-common undoes in advance every closed community. In other words: the fact that we always exist in-common deconstructs a priori every enclosing totality. There is always something and something else, and this is why existence is always shared, plural (see also Lingis; Crim).

We will start with an analysis of Heidegger's ambivalent stance towards the proper and improper. I will analyze Heidegger's line of thought on this issue and how Esposito and Nancy have responded to it. In particular, Nancy's "The Decision of Existence" is of importance here (Nancy, *Birth*), as is Esposito's article on Heidegger and community (Esposito, "Die ursprüngliche Gemeinschaft"), together with, of course, his seminal books *Immunitas* and *Communitas*. I will conclude with an extensive sketch of Nancy's and Esposito's analysis of the intertwining of the proper and improper and the impossibility of abandoning the improper in a proper way.

2 time

I begin with an analysis of Heidegger's concepts of time, Being-with, and thrownness. Each can be read as an attempt to deconstruct the pursuit of a proper being or proper community. Yet Heidegger also abandoned this gesture when he wrote about the "destiny of a people." To understand what is at stake in this "*Kehre*" (turn), we must reflect upon how Heidegger connects identity and community and follow his line of thought as it dispenses with the finite perspective he developed early on, only to later replace it with the appropriation of that which is proper to a certain community.

Let me start with an enigma by way of introducing Heidegger's concept of time: "I am who I am" (Exodus 3.13–15). When asked after the name of his God, this famous answer from Yahweh to Moses is the most condensed expression of undivided properness imaginable. Yahweh is either enclosed upon Himself or can pretend He is because of one condition: as an eternal being, He is immune to time. Yahweh *is*, but He is out of time. He is, so to say, expelled from Being as described in Heidegger's *Being and Time*.

Time, Heidegger argues, confronts us with the question of transcendence – not beyond, but within the world – as a crucial condition of disclosedness. As he observes in §69 of *Being and Time*, when we think about Being and time, not only must we ask about the disclosedness of the world but also about how Dasein ek-sists. How is it possible that something like the disclosedness of the world can belong to the Being of what exists within this world, namely Dasein? In answer to this question, temporality is of the utmost importance, and this is why Yahweh is no Dasein and vice versa. Yahweh exists beyond time and therefore He can appear to be an undivided properness.

We human beings, on the other hand, are destined to be improper and divided. Timeliness (*Zeitlichkeit*) conveys the fact of being open, of reaching out into the world in different directions, of being more than an enclosed entity.[1] Time is the condition of possibility of all the ways in which Dasein concretely exists (Raffoul, *Heidegger*; Dastur). The temporal dynamic that fundamentally marks Dasein, as termed by Heidegger, has three ecstasies, namely: the future, the present, and the past. This is why timeliness is itself ek-static rather than a substantially unified time in which the three ecstasies or temporal dimensions coincide (Heidegger, *Metaphysische*).

As Heidegger describes it in *Being and Time*, ek-static timeliness constitutes the disclosedness (*Erschlossenheit*) towards the world of Dasein. Dasein is never simply present in-itself but is always marked by timeliness and is thus already outside of itself, open towards, and in this sense, transcending towards the world. Dasein is always already transcending, already opened up towards social Being-with. The concept "ontological transcendence" describes this being-in-the-world (Raffoul, "Otherness" and *Heidegger*). Dasein is already in a world, and the world is not encountered in and as a collection of objects as it is conceived from the perspective of dualism between a subject and an object. The subject is no longer the underlying, self-identical unity that remains unaltered and unaffected by all its encounters. Ecstatic timeliness, the finitude of our being-in-the-world, is what makes the appearance of such a subject impossible. We are already in-the-world and such an ontological transcendence undoes in advance any claim to auto-productive individuality or an immanentist, "proper" identity (Heidegger, *Metaphysische*).

From this analysis, Heidegger seems the perfect catalyst for thinking the improper of existence and the impossibility of separating the proper from the improper. For Heidegger, time is a concept that makes a "proper" life impossible for humans, and therefore there is no suggestion of abandoning the improper. One would presume that the same logic applies to community or Being-with: neither individual nor community can attain an originally proper identity fully present to itself, as proper and improper are intertwined in the category of Man.

And yet, in Heidegger, we find at least two ways of thinking about the improper. On the one hand, according to the analysis of time, the improper is something we can and should not abandon. On the other, the same Heidegger falls back upon the thought-schema of an original, proper, or authentic community that must be recovered (Heidegger, *Being and Time*). With this claim, Heidegger leaves behind his own analysis of the interplay of proper and improper which constitutes the existential category of Man; Man which is destined to an existence in which proper and improper are intertwined with each other.

Only when Heidegger "forgets" the impossibility of abandoning the improper in a proper way can he opt for thinking community as the destiny of a particular people with a shared, proper identity which leaves the improper behind. Considered in this way, it is as if the closed entity of the community would be immune to the improper, thereby contradicting his own existential analysis and the double bind of proper and improper. From this perspective, Being-with is no longer a matter of the Being of community but of the community as a given, which eventuates in its ownmost way and belongs to our Being as a proprium that can be appropriated. That is to say, as found in Nancy's analysis, "being" is substituted for "having" (Nancy, *Sense*). Before I present Nancy's analysis, and in order to understand what is at stake in Heidegger's oeuvre as such, I need to further analyze Heidegger's concept of community and Being-with.

3 community

With (the early) Heidegger, the return to an original proprium or the abandonment of the improper is impossible because the origin and ground of the community have always already fallen away from pure presence. The finite condition of our existence is that we are thrown into Being-there: "As thrown, it [Dasein] has been submitted to a 'world,' and exists factically with Others" (Heidegger, *Being and Time* 435). This thrownness, then, is not to be conceived of as a fall from a previous situation of fullness. The ineradicable groundlessness of thrownness, the very fact of existing without ground – rather than the abyssal nature of thrownness, which could indulge yet another lament about the loss of a prior foundation – is precisely what excludes such a situation of original fullness. Whereas the abyssal character of thrownness might still indicate, in a negative way, a previously grounded state, groundlessness means that we exist without any foundation or essence. This essenceless existence that we have to be is what we share but can never own as is if it were our property. By approaching the question of community from this position of ontological transcendence, Heidegger avoids the logic of striving to recover a lost proprium, which is always on the verge of reinstating social disintegration.

Heidegger does not pay much attention to community in his fundamental ontology. Moreover, several commentators have lamented that his attention to the Being of the "we" is one-sided and that it remains limited to a negative theory of community.[2] For Heidegger, it is only the assumption of one's ownmost and individual existence – the "each case mine" (*Jemeinigkeit*) that I will have to be – that constitutes one's proper existence. It is this that has incited some to detect yet another version of the self-present subject in Heidegger's work (De Fontenay; Farias; Olafson; Polt). Heidegger's critique of our everyday preoccupations with the other in the order of "the They" would then merely point to another avatar of originary thinking and of property.

However, one can argue with Heidegger's writings that *Jemeinigkeit* can never be understood as an individualistic self or a self-present individual (Raffoul, "Otherness"). Heidegger's critique of a certain form of Being-with, namely, the absorption of individuals in "the They," derives from his reservations about an identification-based being-together. "The They" is just such a being-together. In the public realm of "the They," all forms of alterity are undone because self and other are sacrificed – "leveled," writes Heidegger – to the same. Every Dasein and all otherness disappears into an undifferentiated "them," leading to a homogeneous or immanent community:

> Distantiality, averageness, and leveling down, as ways of Being for the "they," constitute what we know as "publicness" [*die Offentlichkeit*]. Publicness proximally controls every way in which the world and Dasein get interpreted, and it is always right – not because there is some distinctive and primary relationship-of-Being in which it is related to "Things," or because it avails itself of some transparency on the part of Dasein which it has explicitly appropriated, but because it is insensitive to every difference of level and of genuineness and thus never gets to the "heart of the matter" [*auf die Sachen*]. By publicness everything gets obscured, and what has thus been covered up gets passed off as something familiar and accessible to everyone. (Heidegger, *Being and Time* 165)

If distantiality, averageness, and leveling down are ways of Being for the They, publicness is articulated as a clearly negative encounter. The They is what undoes all differences between people, which seems to suggest that it only relates to a kind of "superficial" existence. The They echoes some sort of alienated being where all things are leveled.

Precisely because it works against such a leveling process, Heidegger's Being-with cannot be assimilated to any homogeneous community. The criticism that Heidegger's notion of Being-with is expressed in a merely negative way – that is, that Heidegger delineated Being-with only on the level of the "improper" and that the corresponding indeterminate everydayness, to which all of us belong and which has as at its *subiectum* (in the sense of "underlying") "the They" – neglects the fact that, for Heidegger, the order of the proper and improper are intimately connected with one another. "The They," for Heidegger, is an existential, and therefore part of Dasein's ontological make-up: one cannot escape it and its everydayness. As a result, the difficulty (or discussion) of interpretation lies in the fact that Heidegger makes no moral judgment about the They – which is very similar to the way that the proper and improper are not to be interpreted as moral categories – and this is where Roberto Esposito's analysis enters the scene: proper and improper always already occur simultaneously, Esposito argues. As a result, in Heidegger, the individual or the community never aim at realizing an actual essence. The sole "essence" at our disposal is our essenceless existence. The proper, as

Esposito concludes, exists in Heidegger solely as the consciousness of our own improper (Esposito, *Communitas*). To strive for an authentic community has no place in such logic.

4 the destiny of a people

Heidegger, it seems, offers us a well-articulated and detailed framework with which to think the improper character of our being. However, community and property in Heidegger's oeuvre become entirely different questions, as can be seen in that famous and often-discussed passage in Part II of *Being and Time* where Heidegger describes the destiny of a people as a *Geschick* into which the individual seems to disappear:

> But if fateful Dasein, as Being-in-the-world, exists essentially in Being-with Others, its historizing is a co-historizing and is determinative for it as destiny [*Geschick*]. This is how we designate the historizing of the community, of a people. Destiny is not something that puts itself together out of individual fates, any more than Being-with-one-another can be conceived as the occurring together of several Subjects. Our fates have already been guided in advance, in our Being with one another in the same world and in our resoluteness for definite possibilities. Only in communicating and in struggling does the power of destiny become free. Dasein's fateful destiny in and with its "generation" goes to make up the full authentic historizing of Dasein. (Heidegger, *Being and Time* 436)

To understand what is at stake in this passage, we must reflect upon how Heidegger makes the connection between identity and community. In Heidegger, the community is something we will have to be – a finite condition that constitutes me rather than the silhouette of a lost or yet-to-be-attained commonality. Since *Dasein* is also *Mitsein*, every mode of Being is always a finite being.

This same Heidegger dispenses with that finite perspective in order to replace it with the previously explained logic of the double bind in which the category of the proper was understood as but the other side of the improper. At one point, Heidegger again speaks of the appropriation of that which is proper to a certain community. From this perspective, it is no longer a matter of the *Being* of community but of the community as a given which eventuates in its ownmost way and belongs to our being as a proprium that can be appropriated. That is to say, *to be* is substituted with *to have*.

This "*Kehre*" in Heidegger is of equal importance to his own stipulated *Kehre* in his thought. Nothing less than the existential analysis and the question of being is put aside as if these were mere lifestyle preferences. Esposito puts his finger on Heidegger's sore spot:

> Community – once understood as something which precedes our condition – has to be recognized in our destination and, consequently, needs to be constructed and reconstructed in its original essence. (Esposito, *Communitas* 127; my translation)

Esposito argues that this is the horrible syllogism that imprisons Heidegger in his own discourse. Although Heidegger offers us a rigorous thinking of community, by linking community with destination it is transformed into a kind of "originary thinking" and, from a political point of view, the striving of one particular community to realize its destiny, which is to recapture its very origin.

It is at this moment that Heidegger attempts to bring Being-with under the category of a property to be appropriated and as a destiny worth striving for by substituting a finite Being-with for a thought of origin and the "*Geschick des Volkes*," the destiny of a people. The evidence of this lies in the passage cited above from *Being and Time* on the fate of a people. There, the community is the destiny of a very particular community: namely, German society.[3] It is only this community that can realize an actual existence, as Heidegger says in his Rector's address:

> The assumption of the rectorate is the commitment to the spiritual leadership of this institution of higher learning. The following of teachers and students only awakens and strengthens through a true and common

rootedness in the essence of the German University. This essence, however, only gains clarity, rank, and power if the leaders, first and foremost and at any time, are themselves led – led by the relentlessness of that spiritual mission that forces the destiny of the German people into the shape of its history. (Heidegger, *Die Selbstbegauptung* 107)

5 the decision *of* and *for* existence

How should we analyze this problematic? And what can we learn from Heidegger's gesture to understand the improper today? Of the many commentators, Esposito and Nancy offer us perhaps the most detailed analysis of Heidegger's schizophrenic stance towards the improper. Both Nancy and Esposito are crucial for understanding why existence is always proper and improper. I will follow their line of thinking in order to find an answer to the question of the improper today. Esposito states that in order for there to be something common, there must be something different from the proper. In order to live, a community is condemned to include what it wants to exclude. Consequently, we are subjects of nothing but our own expropriation (Esposito, *Immunitas*).

As is the case with Esposito, Nancy's thinking not only helps to understand why the improper is currently of crucial philosophical relevance but it also makes "sense" politically. Nancy starts from the idea of an impossible appropriation of the proper and aspires to think about the improper from an existential analysis of Being and from an analysis of sense. I will conclude with the analysis of sense in Nancy's oeuvre.

Before doing so, however, I would like to discuss in detail what Nancy means when he states that Heidegger is preoccupied with the destiny of the proper because he confuses two "decisions," namely, the decision *of* existence and the decision *for* existence. Nancy states that we are thrown into the world and opened to that world, and that this is never our decision. Furthermore, he argues that we can never appropriate this "original" openness. This means that our existence is constituted by thrownness without orientation, resulting in the undecidability of one's essenceless existence, and by the fact that Being is always at once authentic and improper. Nancy concludes that, having an existence which is "in essence" proper and improper, community can never enclose itself within the proper, as it is always already thrown by a constitutive improperness.

Let me unravel this statement by discussing in detail the line of Nancy's argument. In "The Decision of Existence" (Nancy, *Birth*), he discusses Heidegger's stance from what he calls a "double decision." This double decision includes, firstly, the decision *of* existence (*décision de l'existence*), and, secondly, the decision *for* existence (*décision pour l'existence*). The first decision has to do with the fact that I am thrown into the world and open to that world. This fact or moment is not my decision. We can never appropriate this "original" openness or this onto-*ouverture*: the onto-ouverture can never be an auto-ouverture (ours, mine). The openness of Dasein, that is, its thrownness without orientation and without a why, and the fact of being at once authentic and improper, constitutes the undecidability of one's essenceless existence. This undecidability is existential and happens of its own accord. It is, as it were, a decision made by existence itself.

The second decision, on the other hand, the decision *for* existence, does pertain to us. This consists in assuming the undecidability of one's existence as one's own, that is, as in each case mine. Because we are thrown into the world, we have the possibility of regarding this thrownness as our own and of deciding to take our existence into our own hands, without this decision being attributed to a specific origin of existence (Nancy, *Birth* and *Experience*). The decision for existence is in each case mine.

Heidegger's mistake, according to Nancy, lies in confusing these two decisions. Nancy then tries to make this clear through reference to Heidegger's ambivalent account of "the They." On the one hand, in standard Heideggerianism, the everydayness of "the They" is in no way a mediocre order. It is an existential structure in which we are immersed, whether we like it or not. We cannot substitute "the They" for a more authentic existence for the simple reason that this decision is not up to us. We can only decide to exist, but what is proper to existence cannot be distinguished from "existentiel" existence (Nancy, *Birth*). This is as much as saying that, for Heidegger, existing is always proper and improper. Proper existence can thus never simply be authentic (in a moral sense), a pure or original existence without any reminiscences of the improper of "the They." Improper existence is not an object facing us and which one or another "*Volk*" can appropriate as its own. To abandon the improper of "the They" – which, as such, is not even possible – is thus not the same as appropriating the proprium of a people anew.

According to Nancy, in *Being and Time* there is an unmistakable tone of disapproval concerning "the They," which tries to turn it into an (morally) inauthentic rather than an (neutrally) improper world (Nancy, *Birth*). With his plea for a community of destiny, Heidegger appears to forget his own analysis that the existential structures of proper and improper are always immersed in existential Being, as Nancy writes, and they are not categories we can simply abandon or leave behind. Not only does Heidegger forget the naked and brute fact of being-thrown-into-the-world, he also forgets his earlier logic concerning the simultaneity of proper and improper existence, and of the impossibility of extricating one from the other. Both are indeed poles of the condition of Dasein and, consequently, of the Being-with into which we are thrown. The earlier tension between both existential poles is left behind in favor of a moral preference for an "existentiel" proprium and for a spiritual mission of the German people. In this way, Nancy argues, Heidegger appropriates the decision that, according to his earlier logic in *Being and Time*, could and cannot be appropriated. He nevertheless tries to turn the onto-opening into an auto-opening (Nancy, *Birth*). Put differently, he confuses the decision for existence with the decision of existence.

6 the impossibility of proper abandonment

To summarize Nancy's point: only when Heidegger leaves the intertwining of proper and improper behind is he able to imagine the possibility of an original and authentic moral life, to leave the path of his own existential analysis and revert to an originary logic which was excluded earlier on in *Being and Time*. Given the dramatic consequences of Heidegger's "decision," Nancy not only warns us against the risks of this gesture, he also pleads not to leave the logic of the double bind of proper and improper behind. Or even more exactly, Nancy attempts to demonstrate how one can never "leave" this double bind behind in a proper way. For Nancy, thinking community from and with an existential analysis is to deconstruct the illusion that one can abandon the improper in a "proper" way, meaning that all properness is always related to an outside, to something that does not belong to the proper and contaminates it.

In addition, Esposito offers us conceptual tools to demonstrate this point, albeit from a slightly different angle. In several books and articles he defends the thesis that community undoes the proper. In *Immunitas*, for instance, he develops this line of thought, namely, that in order for there to be something common there must be something different from the proper. In order to live, a community is condemned to include what it wants to exclude. Consequently, we are subjects of nothing but our own expropriation (Neyrat; Esposito, *Immunitas*; for an interesting sketch of "immunology" in Nietzsche, see De Cauwer).

In contrast to Nancy's double-bind logic, Esposito starts from the aporetic dialectic between common and proper, and between immunity and community (Dona). If immunity is the power to preserve life, it also is a destructive power for community. Immunity protects and negates, it is "the intrinsically antinomic mode by which life preserves itself through power. From this perspective, we can say that immunization is a negative [form] of the protection of life" (Esposito, "Immunization" 46).

Within the immunity paradigm, immunity is never simply the negative side of community, but the negative that threatens and constitutes the community. Starting from an etymological reading of *communitas* and *immunitas* (see also Esposito, *Communitas*; Farneti), in his interpretation Esposito describes the origin of community and the common as the end of the proper and my "own." The common is not my own but it is what is collective in contrast to what is particular. Community is giving up the proper (Dillon; Nancy, "Being-With"). Immunity then presupposes community but also negates it. Immunity is what protects community by interjecting what threatens it. In order to survive, to be immune, the community is contaminated by its own outside. In other words, it is only able to function as a proper community by allowing the improper it wants to exclude.

In "Community and Nihilism," a paper from 2009 that largely develops arguments from *Communitas*, Esposito elucidates this point with an analysis of the customary opposition between nihilism and community (Esposito, "Community"; Meloni). Community, Esposito argues, is inhabited by an absence because it is always the community of others and never of oneself, and this means "that its presence is structurally inhabited by an absence – of subjectivity, identity, and property" (Esposito, "Community" 46).

In *Communitas*, Esposito argues, with a philosophical and etymological analysis of "*communitas*" and "*munus*," that community is not linked to a surplus but to an absence, a deficit (Esposito, *Communitas*). What the members of a community "share" is not their common substance but the expropriation of it (Oliva). In *Communitas*, Esposito lays bare how the thinking of community, including the thinking of contemporary neo-communitarian movements, has always denied this dimension, thereby simply presupposing that community and nothing are mere opposites, as if community were the thing (*proprio*) itself, excluding the outside, the nothing (*improprio*) (Campbell; Campbell, Luisetti et al.).

Esposito breaks with the idea of community as a whole filled by itself. The outside does not arise when the common or proper breaks apart, but the outside, the improper, is what holds the common – the proper – together. From the very beginning – which is always already the undoing of an original "proper" beginning – the improper is present right at the heart of the proper. Proper and improper, therefore, are each other's "mothers-in-law," to put it in more lyrical terms; neither can lose the other:

> In other words, community is not proscribed, obscured, or veiled by the nothing: it is constituted by it. This simply means that community is not an entity, nor a collective subject, nor a group of subjects. It is the relation that makes them no longer be individual subjects, since it interrupts their identity with a bar that passes through them and thus changes them. It is the "with," the "between," the threshold on which they cross in a contact that relates them to others to the very extent that it separates them from themselves. (Esposito, "Community" 27)

If the proper can only exist because it is constituted by the improper, not only is the proper – conceived as a whole leaving the improper behind it – an illusion, but also the improper is not something we could simply abandon by means of a voluntary gesture. For Esposito, the co-belonging of thing and nothing, of proper and improper, is a matter of Being, of ontology:

> We could say that community is not the inter of the *esse*, but the *esse* as *inter*; not a relationship that shapes being, but being itself as a relationship. This is an important distinction since it gives us back in the clearest possible way the superimposition of being with the nothing: the being of community is the gap, the spacing that relates us to others in a common non-belonging, a loss of what is one's own which never manages to be added up into a common good. Only lack is common, not possession, property and appropriation. (Ibid.)

To paraphrase Esposito: if only "lack" is common, common is lacking. The lack of a common substance cannot be filled in, nor left behind, without being marked by the traces of what constitutes it; the improper is always present in the proper. Community is therefore, Esposito argues, the exteriorization of the inside (Esposito, "Community"; Campbell, Luisetti et al.). Community is not obscured by the veiling of the nothing: it is constituted by the nothing. There is no "real" possibility lying behind the improper. The opening of the proper can never be filled in, enclosed upon itself or cicatrized without using the cord of the improper. Giving up, as Frederic Neyrat also points out in "The Birth of Immunopolitics," should be understood as "ex-isting, a movement outside of oneself" (Neyrat 33). Living in a community is living outside of oneself and therefore living in the improper. How, then, could we ever leave the improper behind or choose a proper existence, as Heidegger once maintained, if we ourselves are always already left behind by it? As Esposito states, community is never a place of arrival, but always of departure. This departure, however, is never a leaving behind or overcoming.

7 concluding remarks: being and sense, and nothing, and something

As was also the case for Nancy, it is remarkable that Esposito refers to Heidegger to argue his thesis whilst the same Heidegger represents the very idea that Esposito attempts to undermine. Esposito uses Heidegger to suggest "a recommencement of the thought of community which Heidegger and Bataille could only guess, but not thematize" (Esposito, "Community" 34). For Esposito, this direction is one which is able to confront the aporetic co-belonging of proper and improper, of thing (community) and nothing (nihilism), and not one in which the void or absence opposes the idea of a common destination of a people. If we want to develop a contemporary thought on the proper and the improper in relation to community, it can never be situated in the attempt to refill the void that (modern) community confronts us with. On the contrary, we need a thought commencing from the constitutive necessity of the absence "present" in the heart of the presence of our being-together (Radin).

Also, for Esposito, the question is indeed how not to abandon the improper in a proper way. This question has been unraveled by rethinking Heidegger's inability to thematize it. Heidegger's thought on "*das Man*" was, and still is, able to give heed to the co-belonging of proper and improper (Heidegger, *Being and Time*). The constitutive non-authentic life of man is what *das Man* is all about; consequently, since existence is "in essence" proper and improper, community can never consist in enclosing oneself within the proper as it is always already thrown by a constitutive improperness.

Only by forgetting this is it possible to interpret Heidegger as arguing for moral authenticity which can be appropriated in and through a collectivity. Heidegger indeed thought that the National Socialist movement would guarantee this collective liberation of the improper. But just how this ought to happen was unclear even to Heidegger at the time of writing *Being and Time*. This, however, did not prevent him from performing political actions, from connecting his thought with the National Socialist movement and, in time, from identifying with one of the most murderous political regimes ever.

Nancy and Esposito apparently encourage us to think of the improper not only from an existential analysis of Being but also from an analysis of sense. Concluding his paper on "Community and Nihilism," Esposito explicitly refers to Nancy's *The Sense of the World* as an "alternative" to considering community as in-between thing and nothing, proper and improper. Firstly, Esposito refers to the way in which nihilism understands existence as absented from sense: existence as a meaningless and devastating desert in which man can only deplore the loss of mastery over the overall meaning of our experience. But secondly, this devastation, Esposito argues, also opens the space of the contemporary world and makes sense of the world – sense which is then understood as something with no reference to any sense, or meaning, that transcends it:

> Community is nothing other than the border, or transition, between this immense devastation of sense and the necessity that each singularity, each event, each fragment of existence must be in itself meaningful. It refers back to the character, both singular and plural, of an existence freed from any presupposed, or imposed, or postponed sense; of a world reduced to itself, able to be simply what it is: a planetary world, without direction or cardinal points. A nothing-else-than-world. And it is this nothing in common which is the world that associates us in the condition of exposition to the hardest absence of sense and, at the same time, to the opening of a sense yet to be thought. (Esposito, "Community" 35)

It can be questioned whether this sense is "yet to be thought." Starting in the early 1990s, Nancy continuously undertook such questioning, and *The Sense of the World* is one of many books and articles in which he thematizes our being singular plural in the world as sense. The world can only be sense (and vice versa) if it has no sense (to lose), states Nancy. Sense is there just like the world is there: without a why. In the final analysis, the sense of the world is thus a tautology. The world is structured as sense, and sense is structured as world:

> If we are toward the world, if there is being-toward-the-world in general, that is, if there is world, there is sense. The there is makes sense by itself and as such. We no longer have to do with the question, "why is there something in general?" but with the answer, "there is something, and that alone makes sense." (Nancy, *Sense* 7)

Nancy refers to Nietzsche and his distinction between "*Ungrund*" and "*Abgrund*." "*Abgrund*" is the void or absence, which appears only to be the starting point for a renewed longing to fill this gap. "*Abgrund*," as also Esposito refers to, is nihilism. From this perspective, existence is a desert, completely meaningless as it is absented of sense.

"*Ungrund*," on the contrary, does not refer to a lack of foundation or common propriety but more to the lack of communality altogether. "*Ungrund*" stands for existence itself, representing the fact that we are open to existence and to the world. The idea that we are sense – that there is sense and that we are here – is the radical demand and the consequence of the opened space that the world has become for us (Nancy, *Sense*). This is also what Esposito argues: the world which does not "have" any sense is also the open space which makes sense. Being in the world then comprises the sense of our existence. The world has sense to the extent that it stands in relation to a creator outside the world. Once this relation with a creator falls away, the world no longer has sense: rather, it is sense, a space in which there is nothing to abandon, but all the more to exist:

> If the world is not the work of a God, this is not because there is no God, as if this were an annoying circumstance, a privative condition to which one had to accommodate oneself as best one could. (As if, in the final analysis, the world were not complete, as if the causal or final part of the totality had been simply amputated. Often, atheism has not known how to communicate anything other than this.) But there is no God because there is the world, and because the world is neither a work nor an operation but the space of the "there is," its configuration without a face. (Nancy, *Sense* 236)

If we substitute "atheism" for "the improper," the conclusion may be that the improper has indeed often been no more than the negative side of the proper. Following Nancy and Esposito, if there is no proper, but only a world in which proper and improper co-appear, then we have reached a renewed starting point for thinking community today and tomorrow, which is more than the reverse of something that is no longer possible. Esposito realizes this all too well. In *Immunitas*, we read: "By striving to make community more proper to itself, law necessarily makes it less common" (22). Rather than negating the negation (absence) and staying in the illusion of a proper abandonment of the improper, we should deepen the aporetic co-appearance of proper and improper. And this is yet to be thought. As Esposito stipulated in the dialogue with Nancy quoted at the very beginning of this paper: "Community remains our question." The same can be said of the improper: if, today, the question of the improper still "appeals to us" philosophically and politically, we need to understand why proper existence can never simply be a pure or original existence without any reminiscences of the improper. Proper and improper go hand-in-hand, and Esposito and Nancy can be of help in understanding how we can never abandon the improper in a proper way.

notes

1 Heidegger uses the verb "Ent-spannen" (Heidegger, *Metaphysische*). Nancy elaborates on this in *The Sense of the World*.

2 See, among others, "Il y a donc bien, dans *Être et Temps*, un phénoménologie surtout négative de l'être-avec, laquelle ne débouche à aucun moment sur une phénoménologie positive de la sociabilité politique" (Janicaud, "La Politique introuvable" 346–47). See also Polt; Critchley 198–210; and Olafson.

3 In this way, *Being and Time* cannot be separated from Heidegger's political commitments, as Frederik Olafson seems to be claiming in his study of the Heideggerian Being-with: "How can Heidegger's implication in Nazism dictate a judgment on Being and Time?" (Olafson 14). If the author is saying that Heidegger's Nazism need not be deduced from *Being and Time*, I would agree, for *Being and Time* cannot be described as a crypto-fascist book, as is shown by the introduction to this problem by Polt. It would nevertheless be mistaken to sever Heidegger's commitment from his philosophy. This, in fact, is what Žižek claims: "Heidegger did not engage in the Nazi political project 'in spite of' his ontological philosophical approach, but because of it" (Žižek 14).

bibliography

Campbell, T. "Bios, Immunity, Life – The Thought of Roberto Esposito." *Diacritics* 36.2 (2006): 2–22. Print.

Campbell, T., F. Luisetti, et al. "On Contemporary French and Italian Political Philosophy: An Interview with Roberto Esposito." *Minnesota Review* 75 (2010): 109–18. Print.

Crim, A. "Deathbound Community: What Calls for Speaking? (Alphonso Lingis – The Community of Those Who Have Nothing in Common)." *Research in Phenomenology* 25 (1995): 296–305. Print.

Critchley, S. "With Being-with? Notes on Jean-Luc Nancy's Rewriting of Being and Time." *Phänomenologische Forschungen* 3 (1998): 198–210. Print.

Dastur, F. *Heidegger et la question du temps*. Paris: PUF, 1999. Print.

De Cauwer, S. "Robert Musil's Cultural Diagnostics in the Light of Nietzschean Immunology." *Neophilologus* 96.3 (2012): 411–25. Print.

De Fontenay, E. "Fribourg–Prague–Paris. Comme l'être, la détresse se dit de multiples manières." *Le Messager européen* 1 (1987): 13–75. Print.

Dillon, M. *Politics of Security: Towards a Political Philosophy of Continental Thought*. London and New York: Routledge, 1996. Print.

Dona, M. "Immunity and Negation – On possible Developments of the Theses Outlined in Roberto Esposito's 'Immunitas.'" *Diacritics* 36.2 (2006): 57–69. Print.

Esposito, R. *Communitas. Origine et destin de la communauté, précédé de Conloquium Jean-Luc Nancy*. Paris: PUF, 2000. Print.

Esposito, R. "Community and Nihilism." *Cosmos and History* 5.1 (2009): 24–36. Print.

Esposito, R. *Immunitas: The Protection and Negation of Life*. Cambridge: Polity, 2011. Print.

Esposito, R. "The Immunization Paradigm." *Diacritics* 36.2 (2006): 23–48. Print.

Esposito, R. "Die ursprüngliche Gemeinschaft." *Deutzsche Zeitschrift für Philosophie* 45 (1997): 551–58. Print.

Farias, V. *Heidegger et le nazisme*. Lagrasse: Verdier, 1987. Print.

Farneti, R. "The Immunitary Turn in Current Talk on Biopolitics: On Roberto Esposito's Bios." *Philosophy and Social Criticism* 37.8 (2011): 955–62. Print.

Heidegger, M. *Being and Time*. New York: Harper, 1962. Print.

Heidegger, M. *Metaphysische Anfangsgründe der Logik im Ausgang von Leibniz*. Frankfurt am Main: Klostermann, 1978. Print.

Heidegger, M. *Die Selbstbegauptung der Deutschen Universität. Das Rektorat 1933/34. Tatsachen und Gedanken*. Frankfurt am Main: Klostermann, 1990. Print.

Huang, H.Y. "Risk, Fear and Immunity: Reinventing the Political in the Age of Biopolitics." *Concentric: Literary and Cultural Studies* 37.1 (2011): 43–71. Print.

Janicaud, D. *L'Ombre de cette pense. Heidegger et la question politique*. Grenoble: Million, 1990. Print.

Janicaud, D. "La Politique introuvable." *Phénoménologie et politique*. Mélanges offerts à Jacques Taminiaux. Brussels: Ousia, 1989. 333–62. Print.

Lacoue-Labarthe, P. "Neither an Accident nor a Mistake." *Critical Inquiry* 15.2 (1989): 481–84. Print.

Lingis, A. *The Community of Those Who Have Nothing in Common*. Bloomington and Indianapolis: Indiana UP, 1994. Print.

Meloni, M. "Biopolitics for Philosophers." *Economy and Society* 39.4 (2010): 551–66. Print.

Nancy, J.-L. "The Being-With of Being-There." *Continental Philosophy Review* 41.1 (2008): 1–15. Print.

Nancy, J.-L. *The Birth to Presence*. Trans. Brian Holmes et al. Stanford: Stanford UP, 1993. Print.

Nancy, J.-L. *The Experience of Freedom*. Trans. Bridget McDonald. Foreword Peter Fenves. Stanford: Stanford UP, 1993. Print.

Nancy, J.-L. *The Sense of the World*. Minneapolis: U of Minnesota P, 1998. Print.

Neyrat, F. "The Birth of Immunopolitics." *Parrhesia* 10 (2010): 31–38. Print.

Olafson, F.A. *Heidegger and the Ground of Ethics: A Study of Mitsein*. Cambridge: Cambridge UP, 1998. Print.

Oliva, R.B. "From the Immune Community to the Communitarian Immunity – On the Recent Reflections of Roberto Esposito." *Diacritics* 36.2 (2006): 70–82. Print.

Polt, R. "Heidegger and the Nazis." *TPM Online* 14 (2001). Web. 23 Sept. 2011.

Radin, M.J. *Reinterpreting Property*. Chicago: U of Chicago P, 1993. Print.

Raffoul, F. *Heidegger and the Subject*. New York: Humanity, 1998. Print.

Raffoul, F. "The Logic of the With. On Nancy's *Être singulier pluriel*." *Studies in Practical Philosophy* 1.1 (1999): 36–52. Print.

Raffoul, F. "Otherness and Individuation in Heidegger." *Man and World* 28 (1995): 341–58. Print.

Safranski, R. *Martin Heidegger: Between Good and Evil*. Cambridge: Cambridge UP, 1999. Print.

Žižek, S. *The Ticklish Subject: The Absent Centre of Political Ontology*. London and New York: Verso, 2000. Print.

roberto esposito
translated by zakiya hanafi

COMMUNITY, IMMUNITY, BIOPOLITICS

I

Community, immunity, and biopolitics: what is the relation between these three terms through which my recent work has wound its way? Can they be connected together in a relationship that is more than just a simple series of concepts or lexicons? Not only is this possible, in my view, it is also necessary. Indeed, each of these terms takes on its fullest sense only in relation to the other two. But let us start from a historical given, by briefly recalling the transition that the two semantic categories – first community and then biopolitics – went through in contemporary philosophical debate. In the late 1980s in France and Italy, a discourse on the concept of community took form that was radically deconstructive toward the way the concept-term had been used in twentieth-century philosophy as a whole – first by the German organicist sociology on *Gemeinschaft* (community), then by the various ethics of communication, and finally, by American neocommunitarianism. Despite significant differences, what linked these three conceptions was a tendency – which could be defined as metaphysical – to conceive of community in a substantialist, subjective sense. Community was understood as a substance that connected certain individuals to each other through the sharing of a common identity. Based on this understanding, community seemed to be conceptually linked to the figure of the "proper": whether it was a matter of appropriating what is in common or communicating what is proper, the community was still defined by a mutual belonging. What its members had in common was what was proper to them – that of being proprietors of their commonality.

It was in opposition to this conceptual short-circuit – on the basis of which the common was reversed into its logical opposite, namely, the proper – that a number of books appeared in rapid succession, including Jean-Luc Nancy's *The Inoperative Community*, Maurice Blanchot's *The Unavowable Community*, Giorgio Agamben's *The Coming Community*, and my own *Communitas: The Origin and Destiny of Community*. What brought these works into the same arena was a sort of modification of the previous semantic category; in the sense that, quite literally, instead of community referring to a property or a belonging of its members, it alluded rather to a constitutive alterity that also differentiated community from itself, evacuating it of any identity-making connotations. Instead of being united by a substance or *res* (thing), the individuals of a community as it was defined in these works were connected by a fault line that ran through the individuals and reciprocally contaminated them. Especially in the book by Nancy, who had opened up this perspective along a trajectory strongly marked by Heidegger's *Mitsein* and Bataille's *être avec*, community was not conceived of as that which puts certain individuals into relationship, but rather as the very being of the relationship. To say, as Nancy claimed, that community is not a common "being," but, rather, the being "in common" of an existence that coincides with exposure to otherness, means to do away with all substantialist interpretations, whether particular and universal or subjective and objective in character. However, despite the theoretical fruitfulness of this step, a problem still remained to be solved. By removing community from the horizon of subjectivity, Nancy made its political ramifications extremely difficult to articulate – starting from the obvious difficulty of imagining a politics that stands entirely outside subjectivity – thus retaining it in a necessarily impolitical dimension. For this reason the discourse on community continued to fluctuate between a political approach that ended up being regressive – the one on small homelands of soil and blood – and a theoretically fruitful approach that was inexpressible in political terms. My impression is that at the bottom of this difficulty in expressing the new concept of community in political terms there lay a tendency on the part of its theoreticians, and of Nancy in particular, to look at it from the point of view of the *cum* rather than from that of the *munus*. It was as if the absolute privilege given to the figure of relation, of relationship, ended up eliminating its most important content – the very object of mutual exchange – and then, along with it, its potentially political significance as well.

The personal contribution that I tried to make to the discussion was a genealogical shift toward the origin of the concept. I mean that the idea of community bears within itself the key for escaping its impolitical turn and for regaining a political significance; but only by traveling back through history all the way to its Latin root of *communitas*, and even before that to the term from which this derives, namely, that of *munus*. Starting from this assumption, I set out on an interpretative path that, while motivated by the same need as the French deconstructionists, departed significantly from theirs at least with regard to one point in particular. While assuming the *pars destruens* of their discourse against identity-making communautarisms, and staying within the concept of community, I shifted attention from the sphere of the *cum*, which was the focus of Nancy's analysis, to that of the *munus*, which he had somehow left in the shadows. Its complex, bivalent meaning of "law" and "gift" – and, more specifically, of the law of a unilateral gift to others – allowed me to maintain and even emphasize the expropriative semantic category that had been developed by the deconstructionists: to belong entirely to the originary *communitas* means to give up one's most precious substance, namely, one's individual identity, in a process of gradual opening from self to the other. But at the same time, I allowed myself to take a step forward, or rather sideways, which opened up a possible avenue toward the political sphere.

Central to this passage is the paradigm of immunity, which is difficult to access from the side of the *cum* because it derives its negative or privative meaning specifically from the term *munus*. If *communitas* is what binds its members in a commitment of giving from one to the other, *immunitas*, by contrast, is what unburdens from this burden, what exonerates from this responsibility. In the same way that community refers to something general and open, immunity – or immunization – refers to the privileged particularity of a situation that is defined by being an exception to a common condition. This is evident in legal terms, according to which someone who has immunity – whether parliamentary or diplomatic – is not subject to a jurisdiction that applies to all other citizens in derogation of the common law. But it is equally recognizable in the medical and biological meanings of the term, according to which natural or induced immunization implies the ability of the body, by means of its own antibodies, to resist an infection caused by an external virus. By overlaying the legal and medical semantic fields, one may well conclude that if community breaks down the barriers of individual identity, immunity is the way to rebuild them, in defensive and offensive forms, against any external element that threatens it. This applies to individuals, but also to particular communities, which also tend to be immunized against any foreign element that appears to threaten them from outside. Hence the double bind implicit in immunitary dynamics – typical of modernity and increasingly widespread today in all spheres of individual and collective experience, both real and imaginary. Although immunity is necessary to the preservation of our life, when driven beyond a certain threshold it forces life into a sort of cage where not only our freedom gets lost but also the very meaning of our existence – that opening of existence outside itself that takes the name of *communitas*. This is the contradiction that I have sought to bring to attention in my work: that which protects the body (the individual body, the social body, and the body politic) is at the same time that which impedes its development. It is also that which, beyond a certain threshold, is likely to destroy it. To use the words of Walter Benjamin, one could say that immunization at high doses is the sacrifice of the living – of every qualified life, that is – for the sake of mere survival. It is the reduction of life to its bare biological matter. How the category of community can regain a new political significance, without ending up in a substantialist metaphysics, becomes clear thanks to this hermeneutic key. The moment the immunitary dispositif becomes the syndrome of our time, one that is both defensive and offensive, community presents itself as the chosen locus – the real and symbolic form – of resistance to the excess of immunization that relentlessly entraps us. If immunity tends to shut our existence up into non-communicating circles or enclosures, community is not so much a larger circle that contains them as it is a passage that cuts through their boundary lines and mixes up the human experience, freeing it from its obsession with security.

2

But here is where the second question that we began with gets grafted onto the first: the type of politics that we are speaking about in this case can only be a form of biopolitics. Since the phenomenon of immunity is inscribed precisely at the point of intersection between law and biology, between medical procedure and legal protection, it is clear that the politics that it gives rise to, in the form of action or reaction, must be in direct relationship with biological life. But the relationship between biopolitics on the one hand and the oppositional dialectic between community and immunity on the other is even more intrinsic – because it touches on the otherwise elusive meaning of the various kinds of dynamics bundled together that can all be traced to the biopolitical paradigm. It would be pointless to reconstruct the recent history of this paradigm here, one that originated from courses that Michel Foucault gave in the 1970s and that was pursued mainly by Italian interpreters – initially by Giorgio Agamben and Antonio Negri – who developed the extraordinary insights of the French thinker along original lines.

These authors' different approaches to the category of biopolitics, however, are precisely what point to a difficulty, or better yet, to a fundamental antinomy – one that is somehow recognizable in a latent form in Foucault's works – that consists essentially in a missing, or inadequate, joint between the two poles of *bios* and politics out of which the term "biopolitics" is composed. Rather than being joined into a single semantic block, it's as if they were designed separately and then only later related to each other. What I mean is that the radical difference between a negative, if not apocalyptic, type of interpretation and an opposite, markedly optimistic and even euphoric interpretation of biopolitics is embedded in a semantic breach, already to be found in the writings of Foucault, between two layers of meaning in the concept that were never perfectly integrated and, indeed, that were destined to break it into two parts that are mutually incompatible or only compatible through the violent subjugation of one to the dominion of the other. Thus, either life appears to be seized, and seemingly imprisoned, by a power destined to reduce it to mere organic matter, or politics risks remaining dissolved in the rhythm of a life that is able to reproduce itself without interruption beyond the historical contradictions that assail it. In the first case, the tendency of the biopolitical regime is to not deviate from the sovereign regime, of which it appears to be an internal fold; in the second case, it emancipates itself from the sovereign regime almost to the point of losing all contact with its deep genealogy. As we noted earlier, Foucault himself never came to a decisive choice between these two extreme possibilities, and he continued to vacillate from one to the other without ever arriving at a definitive solution. In his categorial apparatus, the relationship between the sovereign and biopolitical regimes as well as that between modernity and totalitarianism remained clouded by this fundamental indecisiveness regarding the meaning of what he himself called "biopolitics" or "biopower" – without attributing any particular significance to this lexical difference – and even more regarding its outcomes. As I have already observed, my impression is that there is something missing from his formidable conceptual dispositif – a link or a joining segment between the two – that is able to connect these different configurations of the concept and, even prior to that, to connect the two polarities of life and politics into a more organic, complex form than the one he created, still hesitantly, in his pioneering work.

This constitutive nexus is what I have sought to identify in the paradigm of immunization. In its dual appearance in the legal and biological realms, this paradigm is the exact point of tangency between the spheres of life and politics. This is where the possibility arises of filling the gap in principle between the two extreme interpretations of biopolitics – between its deadly version and its euphoric version. Instead of two opposing, irreconcilable ways of understanding the category, they constitute two internal possibilities, in a horizon that is unified precisely by the bivalent character of the immune dispositif, which is both positive and negative, protective and destructive. Once the dual character of the immunization process has been established – as both the protection and negation of life – the paradigm of biopolitics or biopower can also be defined more congruously by taking immunization as a starting point. The negative mode that has characterized the biopolitical paradigm is not a result of the violent subjugation that power exerts on life from the outside, but rather the contradictory way that life itself tries to defend itself from the dangers that threaten it, contradicting its other equally prominent needs. Immunity, necessary to the preservation of individual and collective life – none of us would stay alive without the immune system in our bodies – if assumed in a form that is exclusive and exclusionary toward all other human and environmental alterities, ends up counteracting its own development.

At stake, if you like, is the difference – which Derrida brought to the fore in another fashion – between immunization and self-immunization. We all know what autoimmune diseases are. They are pathological conditions that occur when our body's immune system becomes so strong that it turns against itself, causing the death of the body. This does not happen all the time, of course. Normally the immune system is limited to a role of preservation, without turning against the body that houses it. But when this does happen, it is not provoked by an external cause but rather by the immune mechanism itself, which is intensified to an intolerable degree. A similar dynamic is also recognizable in the body politic, when the protective barriers against the outside begin to represent a greater risk than what they are intended to prevent. As we know, one of our society's greatest risks today lies in an excessive demand for protection, which in some cases tends to produce an impression of danger, whether real or imagined, for the sole purpose of setting up increasingly powerful preventive defense weapons against it. This logical and historical articulation between the paradigms of biopower and immunization allows us, on the one hand, to clarify the meaning of the concept of biopolitics and, on the other, to establish an internal distinction between its negative mode and another potentially affirmative mode. The fact that the first mode was far more prevalent throughout the course of the last century than the second does not mean that the affirmative mode cannot make a reappearance.

But first things first. It has often been questioned whether the category of biopolitics has any true specificity, since politics has always, in one way or another, been involved with life, even in its strictly biological sense. Weren't agrarian policy in ancient Rome and the use of slave bodies in ancient empires forms of biopolitics? So what distinguishes them, in essence, from what has been called biopolitics? Furthermore, did biopolitics come into being with modernity, as Foucault was inclined to believe, or does it have a longer, deeper genealogy? It could be said in answer to these questions that, looked at from the point of view of its living matter, all politics has taken and will take a biopolitical form. But what caused the first modern intensification and, later on, during the totalitarian phase, its thanatopolitical development, was its immunitary character. As Nietzsche saw clearly, what we call "modernity" is nothing but a metalanguage that has allowed us to respond in immunitary terms to a series of preventive security demands that arose from the very depths of life at a time when the promises of transcendental salvation had vanished. If the paradigm of immunization helps us to understand the structural link between modernity and biopolitics, that of autoimmunization allows us to establish the relationship as well as the element of discontinuity between modern biopolitics and the thanatopolitics of the Nazi regime. In the latter instance, not only had the racial defense of the Germanic people become the main focus of German politics – in a form that made their survival conditional on the death of their external and internal enemies – but, at a certain point, when defeat seemed inevitable, the order was even given for self-destruction. In this case, the immune syndrome took on all the lineaments of an autoimmune disorder, and biopolitics came to coincide perfectly with thanatopolitics.

3

As has now become clear, the end of Nazism – and then, nearly half a century later, of Soviet communism – did not mark the end of biopolitics, which has settled permanently in contemporary society in a form that may appear to have replaced the old ideologies. It is not difficult to recognize its growing presence in all areas of domestic and international politics, along a line of growing indistinction between public and private. From the health sector to that of biotechnology, from ethnic issues to the environment, the only source of political legitimacy today seems to be the preservation and implementation of life. It is precisely in this context that the need for an affirmative biopolitics returns with new urgency. This would be something – a horizon of meaning – in which life would no longer be the object but somehow the subject of politics. So what sort of shape would it take? Where would we trace its symptoms? With what objectives? This question, or set of questions, is far from easy to answer. Having experienced negative biopolitics, or even overt thanatopolitics, is not enough in itself to identify its opposite by contrast. You cannot just reverse certain practices with deadly aims into positive ones. What is required is a leap in quality in order to set up the connection between constraints and needs in a totally different way, including the expansion of financial markets and the protection of those who are weakest in social, cultural, and generational terms. In this across-the-board effort, only made possible by a new alliance between national and international politics, between individuals and collectives – including parties and movements – a first point of orientation, one that is not solely theoretical, can be found specifically in the dialectic between community and immunity that we referred to earlier. The task at hand is to overturn in some way – indeed in every way – the balance of power between "common" and "immune"; to separate the immunitary protection of life from its destruction by means of the common; to conceptualize the function of immune systems in a different way, making them into relational filters between inside and outside instead of exclusionary barriers. How? Starting from what assumptions? Using what tools? The problem has to be tackled on two levels: by disabling the apparatuses of negative immunization, and by enabling new spaces of the common.

Let's start from the first point. We have already seen how the abnormal growth in devices of control and subjugation leads to a corresponding decrease in individual and collective freedoms. Dividing barriers, blocks to the circulation of ideas, languages, and information, surveillance mechanisms set up in all the sensitive places: these are increasingly extensive forms of devitalization that, on the one hand, we must remove ourselves from, and, on the other, that we must resist by all legitimate means. This is particularly difficult to do. First, because the contemporary dispositifs – from biometrics taken at border crossings to photoelectric cells that capture us in our every move, to wiretapping that records our words or messages – are also ordered to protect society and ourselves. But there's another, more fundamental reason for the difficulty; namely, because as Foucault has made perfectly clear, the subjectification that gives meaning to our practices always takes place through some form of subjugation – so to escape subjugation always involves a desubjectivizing effect. This is why fleeing the dispositifs, or disabling them, always involves a double outcome – of liberation and isolation, of emancipation and impoverishment. Of course, to live outside the network of the Internet is possible – but at no small cost of alienation and disorientation with respect to the globalized world. What one needs to do, before causing them to shut down, or simply not allowing oneself to be captured by them, is to preventively distinguish between dispositifs of prohibition, dispositifs of control, and dispositifs of subjugation; between systems that facilitate our individual and collective experience, and apparatuses that diminish its vital power. Or even to preserve areas of silence in the midst of communications that are now extended to every moment of our lives.

But this is not enough. This is only the negative aspect – that of individual withdrawal – of a strategy that also has to be played out through positive moves. The untying of the immune bonds must be accompanied by the production of common spaces, spheres, and dimensions, which are increasingly threatened by the interference of their opposite. If you think about the term and the concept of "common," it has three opposites – the concepts of "proper," "private," and "immune" – which differ but converge in their contrastive effect. All three, in different ways, are opposed to the semantics of the common in the differing but converging forms of appropriation, privatization, and immunization. These are the three modes by which social ties are dissolved, but even before them, by which the idea of the "common good" is dissolved as well, continually decreasing in intensity and expanse in a world that nevertheless likes to think of itself as global. For some time now, not only philosophers but also legal theorists have started on the task of reconstituting the semantics of the concept of the common good, which is squeezed between the opposing, specular concepts of the private good and the public good. The legal system that arose in ancient Rome as private law was intended to sanction in a legally codified form the originary appropriation of things; but also the appropriation of certain human beings who were reduced to the status of thing by those who proclaimed themselves to be their owners by force. In the modern world, this dynamic of appropriation has been joined by the making public of goods assigned to the control and enjoyment of state bodies. In this way, the space of the common, which cannot be appropriated by individuals or by the state, has become increasingly thinner until it coincides with the legally undecidable area of the *res nullius*, or "nobody's property." When the general immune mechanism was set in motion, this withdrawal of the common –

under the converging pressure of the proper, the private and the public – became even more sweeping. Immunity has not merely reinforced the boundaries of the proper, it has gradually assailed the sphere of the public as well. It is no coincidence that sovereignty has revealed itself to be the first, fundamental immunitary dispositif, along with the categories of property and freedom, which were also preventively immunized.

In the twilight of early modernity, when these categories came into direct contact with the horizon of biological life, the erosion of the common good – which was everybody's and nobody's, nobody's because it was everybody's – it became even more intense. Environmental resources were the first to be privatized – water, earth, air, mountains, rivers; then the city spaces, public buildings, roads, and cultural assets; and finally intellectual resources, the communication spaces, and information tools. All of this while waiting for organs of biological life to be legally available for sale and purchased by the highest bidder. Modernity – with the invention of the state, the largest political dispositif – had already intended to exclude the common good, everybody's good; or at least it reduced it more and more in favor of a dialectic between private and public designed to progressively occupy the entire social scene. In reading authors like John Locke, or even Hugo Grotius, one sees how they went about theorizing the necessity to break down a world given by God to everybody – in other words, to no one in particular – into what belongs to individual owners and what belongs to the state. For a long period of time, yet to end, the concept of government property, as public property of the state, was not the opposite of private property but a complementary aspect of it. With what we usually define as globalization, this kind of "making public" [*publicizzazione*] of the private is increasingly intertwined with the inverse phenomenon of the privatization of the public in a manner that seems to exhaust, and even exclude something like a common good from the horizon of possibilities. This becomes even more important when, with the biopolitical turn underway, any property – whether material or intellectual, corporeal or technological – directly or indirectly comes to involve the sphere of biological life, including in the latter all the resources of the intellect and language, the symbolic and the imaginary, needs and desires.

Now it is precisely on this terrain that the battle for an affirmative biopolitics must be braved and possibly won. It must start precisely by breaking the vise grip between public and private that threatens to crush the common, by seeking instead to expand the space of the common. The fight that has begun against the planned privatization of water, the battle over energy sources, or the one seeking to re-examine the patents granted by pharmaceutical companies that prevent the distribution of cheaper medicines in the poorer areas of the planet all go in this direction. This is, of course, a difficult battle – also because we must not make the strategic mistake of abandoning the public space in favor of the common space, and by doing so possibly facilitate the privatization process. But we must not confuse the common good with that pertaining to the sovereignty of the state or government departments and agencies, which are in any case regulated by the prior juridical division between public and private. The problem is that there are currently no legal statutes and codes aimed at protecting the common in relation to the private, the proper, and the immune. In fact, coming prior to adequate legislation, we currently lack even a vocabulary to talk about something – the common – that was effectively excluded first from the process of modernization and then from the process of globalization. The common is neither the public – which is dialectically opposed to the private – nor the global, to which the local corresponds. It is something largely unknown, and even refractory, to our conceptual categories, which have long been organized by the general immune dispositif. Yet the challenge for an affirmative biopolitics, *of* life and no longer *on* life, is staked precisely on this possibility: on our capacity, even before acting, to think within this horizon, to think around and even from within the common. This is the direction that has been guiding my work over recent years – including through the category of the "impersonal."

peter gratton

SPINOZA AND THE BIOPOLITICAL ROOTS OF MODERNITY

> This frugal, propertyless life, undermined by illness, this thin, frail body [...] how does one explain the impression they give of being suffused with Life itself, of having a power identical to Life [...] [There] lies the *glory and innocence* of Spinoza, his own discovery. In a world consumed by the negative, he has enough confidence in life, in the power of life, to challenge death, the murderous appetite of men, the rules of good and evil, and the just and the unjust. Enough confidence in life to denounce all the phantoms of the negative.
>
> *Gilles Deleuze,* Spinoza: Practical Philosophy[1]

There is, no doubt, a repetition of Spinozism bolstering neo-Marxian accounts of the political in works ranging from Althusser and Deleuze in the 1960s to the recent materialisms of Elisabeth Grosz and Jane Bennett, to the important political conceptions of the common in the work of Etienne Balibar and Michael Hardt and Antonio Negri, to the thinking of an affirmative biopolitics in Roberto Esposito. The upshot of Spinoza's thinking is that he presents us with a means of denaturalizing political codes, such as liberal discussion of transcendent rights, that are but ruses for power in modernity. Spinoza's last work, *The Political Treatise,* broaches a political immanentism that Hardt and Negri, among many on the political left, valorize. Taking Machiavelli's power politics from *The Prince* and flipping it and cutting of the king's head, if you will, Spinoza declares a democratic *civitas* always extrinsic to any particular state form or state of affairs. Thus the *civitas* always has a revolutionary power, even, or perhaps especially, when the *civitas* is under the sway of ideologies redirecting its power to the furtherance of the state. But my surmise is that this thinking of *civitas* must be confronted with the critiques of the biopolitical found, for example, in Jean-Luc Nancy and Roberto Esposito, even as these thinkers find much to valorize in Spinoza's immanentism. For readers of this journal, these critiques are well known, especially as it is found in Nancy's *Inoperative Community,* where he denounces political immanence:

> Entangled in its meshes, we have wrung for ourselves the phantasms of the lost community [...] What this community has lost – the immanence and the intimacy of a communion – is lost only in the sense that such a "loss" is constitutive of "community" itself. It is not a loss: on the contrary, immanence, if it were to come about, would instantly

> suppress community, or communication, as such [...] [where] there is only the continuous identity of atoms. This is why political or collective enterprises dominated by a will to absolute immanence have as their truth the truth of death. Immanence, communal fusion, contains no other logic than that of the suicide of the community that is governed by it.[2]

But wasn't Spinoza's writing a testimony to a Nietzschean affirmation of life *avant la lettre*? My task here is not to denounce Spinozism *in toto* but rather to show the ways in which his immanentism pulls along a certain biopolitics that developed in early modernity and from which, as he would be the first to say, he could not make his own work extrinsic. And my surmise is that wherever Spinoza's immanentism repeats itself in contemporary discourse, this biopolitics is pulled along as if by a magnet. "For millennia," Foucault famously noted in the first volume of his *History of Sexuality*, "man remained what he was for Aristotle: a living being with the additional capacity for political existence; modern man is an animal whose politics places his existence as a living being in question."[3] This claim is made in a chapter entitled "Right of Death and Power over Life," and mirrors much of what we also find in Hannah Arendt's depiction of the rise of the nation-state in *The Origins of Totalitarianism*.[4] The study of this "biopolitics" – and the security apparatuses meant to immunize the body politic – has been nothing if not lively in recent years. Many of these studies track certain forms of sovereignty through Carl Schmitt's *Political Theology*,[5] which depicts the sovereign as that being, exceptional to the laws, who decides over matters of life and death, friend and enemy.[6] This *transcendent* form of politics – the sovereign is symmetrical, for Schmitt, to the theological role of God *over* nature – is avowedly Hobbesian.[7] But my claim would be that one should pay attention to the rise of national sovereignty in modernity, a point so obvious and apparent that one is perhaps thought to be naïve not to see the sovereign exceptionalism hovering over its surface. To the contrary: the works of Arendt and Foucault, albeit in different ways, trace these now racialized discourses in terms of immanentist forms of sovereignty that have been far from benign in carrying forth the dictates of safety and security. And this remains the central dictum of Spinoza's politics: "The virtue of a state [*imperii*] is its security [*securitas*]," Spinoza writes succinctly at the end of the introduction to his *Political Treatise*.[8] Let me cite this as the unalloyed core of modern politics, despite its denegation under the popular liberal mottos of equality and liberty. What Spinoza marks, while playing the political realist who refuses not to conceive men as he "would like them to be," but merely "as they are,"[9] is the movement of life itself to the center of political *mores*. To put it another way for readers of this theme issue, there is, at its heart, an argument to be had between Nancy's view of communities of "being-with" and Roberto Esposito's repetition of Deleuzian immanence, which calls on him to find the "outlines of an affirmative biopolitics."[10] To put it too quickly, since it is discussed elsewhere in this special issue, Esposito sees a dialectic between *communitas* (that which binds members to reciprocal donation or giving) and *immunitas* (which sets up a defense against any such giving, as in liberal individualism). For Esposito, a negative biopolitics is one that cannot affirm, on his or Deleuze's account, *a* life.[11] These thoughts have their root in Spinozistic immanentism, and thus guide my reasons for turning to Spinoza in full in this essay. It should be noted that Nancy marks his own distance from Deleuzian immanence in several places, most notably in "The Deleuzian Fold of Thought,"[12] where the negative or the "hollowing" of "being's lack of itself [...] (death, time, genesis, and end)" – that is, a "rupturing" – marks a distance from Deleuzian immanence (112–13). This rupturing is at the heart of Nancy's thinking of community in his early work and the self-rupturing of the "truth of democracy" in his writings of the last several years.

Thus there is an under-discussed biopolitics in Spinoza, who found it scandalous that men of reason could offer views of a monarchical

God ruling over the cosmos from above, who first brought down sovereignty from *en haut* to immanence of becoming itself. For Spinoza, God *is* nature, the ultimate power and immanent cause, and just as Spinoza thought it best not to export our political visions onto the cosmos, so too he would use his own metaphysics to think an immanent politics grounded in the productive forms of power that operate more than by the deduction of previous forms of sovereignty. For these and other reasons I'll discuss, we would be better to begin with Spinoza's political theology than that of Schmitt if we are studying the political shift we call "modernism." While translating Hobbes's premises about the nature of man into his own ontology, Spinoza, just as he had done with the Sovereign God with regard to nature, depicts sovereignty (*imperium*) as immanent to the body politic. "There is, then, a philosophy of 'life' in Spinoza," Deleuze writes.[13] Indeed, but the flip side of this contention is that Spinozist vitalism is not extrinsic to the era of the biopolitical, and we should think seriously about whether we wish it to live on.

At least since Foucault, what has been dubbed the study of "biopolitics" has argued that life *as such* is itself a *political* definition, that there is no life *as such* in all its innocence, not least because the category of the "innocent" life is our most politicized category, if one can speak this way. In other words, the question of life is one that is neither benign nor normatively neutral, but engages in a biologicism that has centered our politics for a number of centuries. But when did this politics of life begin? It is certainly productive of all of our norms and political imaginaries: who, but a nihilist, would wish to deliver us from a politics of life? What could be more *natural* than a politics of life, a politics that would defend life *as such*? Do we not begin to open ourselves to the possibility of the worst if we reveal or critique this kernel at the heart of the political? Can the questioning of the value of life *as such* even be done without risking a politics of death? Is this not the moral blackmail that biopolitics has always offered to thought? Who could not, following Spinoza, affirm life *as such*? Who could not, to use the American parlance, be *pro*-life? It's also true that to attack Spinozism is considered on par with wanting to kill innocent life itself – this poor wastrel of a man, this "innocent life," as Deleuze puts it. We philosophers and theorists admire those who have been threatened and excommunicated, those who have done their philosophy under the most arduous and dangerous conditions, and by this move we flatter ourselves to think that we too brave the worst in merely being philosophers. It seems rude to our self-understanding to attack a Socrates or a Spinoza, as if they didn't face enough in their own lives, and especially if that means siding with those reactionary dogmatists who excoriated "Spinozism" for centuries after Spinoza died having left unfinished his chapter on democracy in the *Political Treatise*. And now, centuries later, he is tellingly judged to be an "innocent life," as if life itself were a matter of guilt or innocence, or a matter of those "suffused with life itself" or those who are not – as if one can make any such juridical distinctions independent of a fully biopoliticized logic. This is where, by the way, I would take my distance from Deleuze's thinking of *a* life, which is notably found for him in a Dickens story where a rogue dies, and is no longer *a* life when he is alive and well, and is judged a "disreputable man, a rogue, held in contempt by everyone."[14] Only purely atomized – a man without qualities – can this life be redeemed. Deleuze's other examples, of course, are small children, where "through all their sufferings and weakness [...] are infused with an immanent life that is pure power and even bliss."[15]

But again, who but the most flagrant antirealist would argue that our notions of life are politicized through and through? Isn't the naturalism of Spinoza an antidote to just this sort of post-Foucauldian, supposed "constructivist," thinking? This would be a mistake for two reasons: (1) Spinoza's work evinces no distinction between the political and the natural. As he puts it in the *Political Treatise*, to think the political aside from natural categories would be to think an *imperium* within an *imperium*, which also has the upshot of being premised on the fact that sovereignty (*imperium*) can never be shared, and thus any commonwealth must speak, as Spinoza argues again and again, with "one voice" and think with "one mind," as if such a thing were possible or desirable.[16] My concern deepens when one considers that, for Spinoza, this speaking with "one voice" is not a *telos* or regulative ideal, since his metaphysics refutes such notions; this "single voice" must follow from the *work of nature* itself immanently unfolding. (2) Spinoza himself argues that the power of the state is productive of the forces of life itself, that is, the very *conatus* that is the essence of the living being in the *Ethics*, while pronouncing what his liberal readers (wrongly) take as his unconditional affirmation of freedom of thought.

Let's zero in further on Spinoza's ontology of power. Spinoza's political claims derive from his *Ethics*, where he argues that there is no other world, no other plane of existence than nature itself. That is, Spinoza's philosophy is engaged in thinking a *pure* immanence, as Deleuze rightly describes. Being "is" univocal; there is no neo-Platonic "good beyond being," which in a sense also means that for Spinoza there is no "good" (neither as *telos* nor as a transcendent, let alone transcendental, value beyond the given concretion of acquired powers). On the one hand, there is no ontico-ontological difference in Spinoza,[17] since all entities are but manifestations of this immanent ground through the various modes and attributes; this would be substance as *natura naturans*, which can be thought as the power productive of any given entity, whether in the mode of thought or in the mode of extended things. This is what is available to philosophical wisdom reflecting on the laws of nature. On the other hand, there is substance as *natura naturata* ("nature natured"), the flow of becoming that is the manifestation of this all-powerful *natura naturans*. Thus one can study nature, understood in its widest sense, either in terms of its unity (*natura naturans*), which requires philosophical wisdom, or in terms of its many, multifarious effects, that is, power as it has instantiated itself, which, for example, is what politicians are particularly good at doing, as Spinoza argues at the beginning of the *Political Treatise*. For Deleuze, Spinoza evinces a trinitary structure: the substance that expresses itself, that particular attribute which expresses, and that mode which is expressed. This is also the place to mention what is the crucial difference between Deleuze and his former student Negri. Where Negri emphasizes Spinoza as a thinker of Being, as a thinker of ultimate homogeneity and unity, Deleuze thinks Spinoza goes too far in this direction: his thinking privileges *the* substance "behind" its manifestations and thus betrays what Deleuze takes to be the principle of his immanentism, namely that Being "is" "equally present in all beings."[18] That is, "Spinoza's substance appears independent of the modes, while the modes are dependent on substance, as though on something other than themselves."[19] In simpler terms, this is why Deleuze is a thinker of becoming.

Politically speaking, for Spinoza his metaphysics means we must deal with the world as it is, not as we would wish it to be – there is simply no other plane of possible existence to which we would compare our contemporary situation. For any political activist this is an important statement of advice – not to let the (transcendental) best be the enemy of the good that may be attained *now*. Spinoza turns Platonism on its head, but in more than just the sense usually understood in terms of his thoroughgoing monism. While one could certainly point out how Plato's *Republic*, *Statesman*, and the *Laws* minutely regulate the living and dying of its citizenry, the *telos* of this *polis* was for the sake of living well, not merely existing as such. The philosopher king required bravery in order to enter the affairs of the state, where it was understood that other principles prevailed. But for Spinoza, *living itself is its own end.* At this point, it's worth recalling his view of life *as such*. "By life," he writes in his early work on Descartes, "we for our part understand the force through which things persevere in their own being."[20] In fact, the negation of death, as Hegel argues in the *Logic*, is displaced from Spinoza's system, since as Spinoza puts it, the "free man thinks of death least of all things, and his wisdom is a meditation of life, not of death."[21] The primary virtue – in all senses, since he notes that no other is "prior to this one" – is for the "*conatus* to preserve itself,"[22] since "nobody can desire to be happy, to do well and to live well without at the same time desiring to be, to do, and to live; that is, to actually exist."[23] Where Aristotle left the necessities of the body to the *oikos*, which was the realm not just of necessity but also of sovereign mastery (over women and children), Spinoza posits that politics comes about precisely as a necessary supplement to the needs of the *conatus* and its autoaffective desire for its own existence. Where Aristotle gives us a politics of freedom and equality (for the few) in the political space, for Spinoza, this freedom must be completely privatized, and the best one can say is that his power politics provides a freedom of thought, but not action, and thus would be an indirect critique of all liberalism to follow: provisions by the state for masses to engage in idle chatter that would have no effect on the functioning of the state. Where for Aristotle complete virtue was for the sake not just of living but of living well (*to eu zen*),[24] Spinoza finds "to act in absolute conformity with virtue is nothing else in us but to act, to live, to preserve one's own being (these three mean the same)."[25] What Aristotle calls *praxis*, Spinoza definitionally reduces simply to continuing one's existence. Living cannot be for the sake of anything else: "nobody endeavors to preserve his being for the sake of some other thing."[26] Thus his politics will be about the protection of "innocent life," that is, life unconditioned by want of anything more, *a* life, as Deleuze puts it, and thus we have the modern claim that "innocent life" is to be cherished, but the "guilty" lives of others are condemned by the same sovereignty that was to protect nothing other than this bare life. Even those often deemed "post-metaphysical," as we've noted, carry this vitalism forward: what else are we to make of Deleuze's conception of the "indefinite life," that is, a "purely immanent life," the primary examples of which – as always isn't it? – are "small children [...] infused with an immanent life that is pure power and even bliss."[27] How many theories of life demand this theoretical labor of children? One can never grow up too late, since childhood is "*complete* power, *complete* bliss." But *pure* immanence, if there is such a thing, is nothing but death, not life, since it would be a black hole from which nothing returns. "*A* life," to use Deleuze's phrase, as pure power, would be nothing of the kind, as Nancy suggests in a different context above in *The Inoperative Community* and, in a still different way, his essay "The Deleuzian Fold of Thought" (1996).

Returning to Spinoza, "life itself" is for him the hidden *natura naturans*, the mystical power that our political modes merely manifest as this or that mode of politics (aristocracy, democracy, or monarchy). Readers such as Negri valorize this aspect of Spinoza, since it means that there is an always-prior "natural" *civitas* or commonwealth that can enact different political modes to which it is irreducible (i.e., there is always the power of revolt). But Spinoza's early readers didn't miss the import of this thinking of the common ground of the *civitas* for the incipient nationalisms of Europe, for which any given state is but a mode through which an original nation enacts itself. Spinoza's "*civitas*" has long been translated in French as "nation,"[28] and his first French translator was none other than the Cont Henri de Boulainvilliers, who, as both Foucault and Arendt point out at length, in the early eighteenth century played a crucial role in the theoretical development of nationalism and the historical search for origins.[29]

Spinoza's thinking requires, then, that the state be nothing other than the protector of the body politic. It is well known that in Hobbes, fear is the political passion par excellence, and that the task of the social contract is to move the individual from a state of generalized fear to the singular fear of the sovereign, who has the monopoly of producing awe in his subjects. As Spinoza himself puts it, "the main difference" between the state of nature and civil society "is this, that in the civil order all men fear the same things and *all have the same ground of security, the same way of life*."[30] But this fear is the flip side of a politics of security, since the contract is for the "*perpetual* (and not temporary) *security*"[31]: "the motive and end [of the social contract] is *nothing else* but the security of a man's person, in his life and in the means of so preserving life as not to be weary of it."[32] For these reasons, one finds in Spinoza's text multiple references to the state's right (i.e., power mitigated by no other authority than carrying out the dictates of this inner principle of the whole) to "compel" enemies "by whatever means,"[33] which, as he puts it, is "arduous work":

> Anyone with any experience of the capricious mind of the multitude almost despairs of it, as it is governed not by reason but by passion alone, it is precipitate in everything [...] [E]veryone knows what wrongdoing people are often moved to commit because they cannot stand their present situation and desire a major upheaval, how blind anger and resentment of their poverty prompt men to act, and how much these things occupy and agitate their minds. To anticipate all this and construct a state that affords no opportunity for trouble-making, to organize everything in such a way that each person, of whatever character, prefers public right to private advantage, *this is the real task, this the arduous work*. The necessity for this has compelled people to seek many stratagems. But they have never succeeded in devising a form of government that was not *in greater danger from its own citizens than from foreign foes* and which was not more fearful of the former than of the latter.[34]

Or, as he argues more strongly in the *Political Treatise*, "It is also beyond doubt that a commonwealth is always in greater danger *from its citizens* than its enemies."[35] We will come back to this "what everyone knows," this shared view of what is shared by every commonwealth, even if it can't be shared, precisely since it's what is beyond doubt and one never has to discuss it. The biopolitical paradoxically presupposes both a literally *innate* living origin (the "same *ground* of security") that the state is to protect (in terms of enforcing the "same way of life") *and* that the threat itself comes from within the nation; it is as if *natura naturata* is always turning against its *natura naturans*, and the sovereign exception is the supplemental force meant to protect the state, to immunize the body politic that threatens it *from within*. Thus nationalist conceptions of homogeneity produce never-ending prostheses of police apparatuses for the safety and security of the multitude that never adheres enough to its living principle. As we will come to it, this is the reason why immanent sovereigntism (Spinoza) and transcendental sovereigntism (Hobbes) are the biopolitical doubles of one another, each supplementing the other in a deadly interplay meant to protect nothing but the life of the body politic, since life itself produces its supplemental prosthesis that is never done immunizing the common, since, of course, *natura naturata* is caught up in a perpetual motion – the "perpetual security" of which Spinoza speaks – that can't reach and secure the inner principle of the political's *natura naturans*. Put otherwise, politics for Spinoza is always already fallen from its original substance and, despite his stated respect for those politicians who enact the political, it is philosophy's wisdom to reflect wholly on the original substance and the origin of the political that is always then betrayed by its material conditions.

Since "what everyone knows" is always already presumed, always already known, we are beginning to see a step away from Hobbes. The basis for later liberalism is Hobbes's discussion of the "authorization" by which we always already accede to the whim of the sovereign,[36] since we would never decide to return to the state of nature. And it's also the case that he opens a space for market liberalism: "the passions that incline men to peace are fear of death, desire for such things as are necessary to commodious living, and a *hope* by their industry to obtain them."[37] At this point, it would be useful to point out that the secondary literature does not take for granted that the *Political Treatise* and the *Theological-Political Treatise* offer visions of the same underlying political theory, with the latter notably leaving out, at least explicitly, a discussion of a

Hobbesian-type contract. This, at least, is the claim of Negri, who champions reading the latter text in terms of the writings of Spinoza's predecessor Althusius. For Negri, this is important, since it doesn't evince the "negative and dialectical essence of contractarianism," which is also the terms he uses when critiquing a tradition that runs through Hegel and Heidegger. What is missing in Negri is that, if indeed Spinoza is "rejecting the contractarian thematic," it is in the name of *an already constituted* (and thus never "constituted" as such) body politic whose sovereign power and vitality is immanent to it. As Negri himself puts it

> the *absolutum imperium* [...] will become a term that, in signifying the *unity* of power, will have to assume it as the projection of the *potentiae* of subjects and to define its totality *as life*, as the always open, internal, dynamic articulation of an organic whole.[38]

The gain found by Negri is that, unlike other forms of government, the democratic unity is taken by Spinoza as a given, natural fact that is an "autonomous productive force" wherein the "power of being thus manifests itself in all its unifying force." This is precisely what one finds in all thinkers of nationalism, from Boulainvilliers and Sieyès onwards.

The crucial distinction between Hobbes and Spinoza, then, is to be found elsewhere. Hobbes's account centered on a multitude, in the classic sense,[39] when he discusses the *salus populi*. Let's see if we can spot the difference. First, let's take a canonical moment from the *Leviathan*:

> A multitude of men are made *one* person [Hobbes's emphasis], when they are by one man or one person represented so that it be done with the consent of every one of that multitude in particular. [In the original Latin, the latter is rendered: "it is the represented who has authority from each one."] For it is the *unity* of the representer, not the *unity* of the represented, that maketh the person *one*. And it is the representer that beareth the person, and but one person, and *unity* cannot otherwise be understood in multitude. And because the multitude naturally is not *one*, but *many*, they cannot be understood for one, but many, authors of everything their representative saith or doth in their name [...][40]

Now, let's pull together several important moments in Spinoza:

> Human society can thus be formed without any alienation of natural right, and the contract can be preserved in its entirety with complete fidelity, only if every person transfers all the power they possess to society, i.e., supreme power, which all must obey, either of their own free will or through fear of the ultimate punishment. The right of such a society is called democracy. Democracy is therefore properly defined as a united gathering of people which collectively has the sovereign right to do all that it has the power to do. It follows that the sovereign power is bound by no law and everyone is obliged to obey it in all things.[41]

Here we seem to find an advance politically over Hobbes, and Spinoza is perhaps the first philosopher in modernity to argue for democracy, which for him is a "state and government where the *safety of the whole people*, not that of the ruler, is the supreme law."[42] Spinoza moves a step further – let's recall we've already seen that all citizens must "live in the same manner" – arguing that the commonwealth must be thought to be an immanent whole. He writes:

> We see, then, that the individual [qua mode] is not in control of his own right [i.e., power], but is subject to the right [i.e., power] of the commonwealth whose every command he is bound to carry out, and he does not have any right [power] to decide what is fair or unfair, what is righteous or unrighteous [...] On the contrary, since the body of the state [*imperium*, which is also the word used by Hobbes for the sovereign] must be guided *as if* [*tanquam*] by a single mind (and consequently the will of the commonwealth must be regarded as the will of all), what the commonwealth decides to be just and good must be held to be so decided by every citizen.[43]

Spinoza repeats this formulation ("as if by one mind"[44] or "as it were, by one mind [*una veluti mente*]") throughout the *Political Treatise* and in his discussion of the Hebraic state in the *Theological-Political Treatise*. At one point, he inserts "*tanquam*" in a quotation from Tacitus about the "one mind" of the body politic – thus, it was clearly a considered phrasing in his texts. This hesitation of the "as it were" or "as if" before proclaiming that one mind is necessary for this body politic, which has but one "purpose [...] which is nothing other than peace and security of life,"[45] may be read as decisive pulling back from thinking the common as always already constituted, as truly acting by one mind and thus, for Spinoza, in one body. But to read it this way would be to misunderstand Spinoza's metaphysical politics, since the body or mind are but attributes of an underlying substance; hence it would be a category mistake to think the *civitas* as being united in mind, since the mind is the effect of the underlying power, not the power itself; thus the "as it were one mind" is the "expression" of the substance's move from *potentiae* to act; the mind is the intermediary between the substance and its particular instantiation in this or that mode, this or that state apparatus. "The right of the state or of the sovereign [*imperii seu summarum potestatum ius*]," Spinoza writes, "is nothing more than the right of Nature itself [*ipsum naturae ius*] and is determined by the power not of each individual but of a people [*multitudinis*] which is guided *as if* by one mind [*una veluti mente*],"[46] and accordingly the task is to "bring everyone's resources, *if we can call it that*, into one body, the body of a society."[47] Thus, for Alexandre Matheron, "Political society is not created by a contract; it is engendered and reengendered at each moment by a consensus that must be permanently renewed."[48] However, this bespeaks not a vital politics but the emptying out of the political as such, which must follow from the laws of necessity, that is, the immanent unfolding of the national substance, wherein all "must emerge from properties that the multitude has in common."[49]

> [E]ither there is no sovereignty nor any right over subjects or else sovereignty must necessarily extend to everything that might be effective in inducing men to submit to it. Thus whatever a subject does that complies with the sovereign's commands, whether elicited by love or forced by fear [...] he still does so by right of the sovereign, not his own [...] And while it is impossible, *of course* [my emphasis], to control people's minds to the same extent as their tongues, still minds too are to some degree subject to the sovereign power, which has various ways to ensure that a very large part of the people believes, loves, hates, etc., what the sovereign wants them to. Our conclusion [is that], without logical contradiction, we can conceive of men who believe, love, hate, despise, or exhibit *any passion whatever*, owing to the power of the state alone.[50]

Needless to say, this "passion" is what, in Spinoza, is central to life itself, its essence, and we will soon see how this will drive his bio-patriarchal politics. Here we see the circularity of Spinoza's biopolitics: the state is the *effect* of the common substance, yet it must *constantly* and *ceaselessly* "work" to produce the essence of the "affects" whose effect is it was already supposed to be. For Negri, the democratic "substance" (i.e., the common), as prior to any state, means that "[d]emocracy is the highest form in which society is expressed,"[51] though one wonders how this or that "form" can make sense as "higher" or "lower" in a Spinozistic conception. In any event, "it is the *highest* form in which society is expressed because it is the most expansive form in which natural society is expressed as political society [...] It is in positive terms the liberation of *all* [my emphasis] social energies in a general *conatus* of the organization of the freedom of all. Continual and permanent."[52] And this political society, as in a state of emergency, may require a dictatorship, Negri writes, which may "tend towards absolutism" and a state of war, which "in this case can transform the possible state of war into an *organic movement* of refoundation and thereby restore the *vigor* to the State."[53] Despite himself, Negri pronounces the transcendental/immanentist logic of the biopolitical, that process of the incessant need of the living thing to "preserve itself," to renew its vitality and vigor, to conquer those threats internal and external. There is thus never a "pure immanence," never a "pure life," since it always requires the prosthesis of the state that will police the health and safety of the body politic, all while producing the very *conatus* that was to be its driving force. Alas, the dead bastards that we would ostracize from our communities, as in Deleuze's story taken from Dickens to represent pure life, have a way of coming back to life, to communities that would wish something other than differentiation and difference.

This brings us to the most perplexing reason to see such a political immanentism heralded by post-Althusserian Marxists, such as Étienne Balibar and Antonio Negri. To the revolutionary dictum that "another world is possible" – that is something beyond the event of a Nancy-an "rupture" – Spinoza's reply is that this is the only world there is, that it necessarily flows from an all-powerful immanent cause, and that the only wisdom one can have is to reflect on God, that is, Nature, in terms of its all-encompassing causality. This neo-stoicism, then, will herald a completely privatized freedom of thought, while all action is to flow from the necessity of the power of the state and its tending to the safety and protection of life. One could see in Spinoza, then, a wonderful passkey to the liberalism that was to follow: the apparatus of the state with regard to its legitimacy in matters of life and death is to remain unquestioned, all as speech is to be freed in matters of private reason: "freedom of spirit or strength of mind is the virtue of a *private* citizen."[54] In sum, we have the charade that centuries later Arendt would find in Sartre's depiction of one being free even in a prison – an absurdity to those who think freedom is something other than that which occurs in the privacy of one's body or mind, unrelated to a web of relations with others. It is no wonder that the best Negri can offer, for example, is that his Spinozism is based on "try[ing] to understand how the concept of democracy *could have been expressed* in the *TP*."[55]

But we do not have to travel into the biopolitical future after Spinoza to see the status of this biopolitics in his text. As always, the biopolitical never neglects the role of women and reproduction (even if it's notable that in the work of Foucault, Esposito, Agamben, and Negri the biopolitics of women is simply ignored or denegated), and we don't need to engage in the necrophilia of fantasies about what Spinoza *would have* written if only he hadn't died. Here we see how biopolitics founds all political naturalisms, which, as in Spinoza, forms a determined, causal relation between nature and the political, between *natura naturans* and *natura naturata*:

> But, perhaps, someone will ask, whether women are under men's authority by nature or institution [*ex natura an ex instituto sub potestate virorum sint*]? For if it has been by mere institution, then we had no reason compelling us to exclude women from government. But if we consult experience itself, we shall find that the origin of it is in their weakness [*imbecillitate*]. For there *has never been a case of men and women reigning together* [my emphasis], but wherever on the earth men are found, there we see that men rule, and women are ruled, and that on this plan, *both sexes live in harmony* [*hac ratione utrumque sexum concorditer vivere*] [...] But if by nature women were equal to men, and were equally distinguished by force of character and ability, in which human power and therefore human right chiefly consist; surely among nations so many and different some would be found, where both sexes rule alike, and others, where men are ruled by women, and so brought up, that they can make less use of their abilities. And since *this is nowhere the case*, one may assert with *perfect propriety*, that women have not by nature equal right with men: but that they necessarily give way to men, and that thus it cannot happen, that both sexes should rule alike, much less that men should be ruled by women. But if we further reflect upon human passions, how men, in fact, generally love women *merely from the passion of lust*, and esteem their cleverness and wisdom in proportion to the excellence of their beauty, and also how very ill-disposed men are to

> suffer the women they love to show any sort of favor to others, and other facts of this kind, we shall easily see that men and women cannot rule alike without great hurt to peace. But of this enough [*Sed de his satis*].[56]

Hasana Sharp rightly asks:

> If Spinoza's concern throughout his *Political Treatise* is to institute the most robust and harmonious form of sovereignty, in which the commonwealth is guided as much as possible by "one mind," why exclude a vast majority of the population from the business of running the state and determining its representatives?[57]

Sharp notes a whole literature in recent years that finds Spinoza's views on women "surprising." My take, however, is that one cannot excise the "blank page" (the name given to his last written words) as a contingent feature of Spinoza's system, not least because of the irony that Spinoza's system evinces no concept of the "surprise" or, indeed, contingency itself – everything is in determined cause-and-effect relations. And on each point one can find an isonomous description between this depiction of women and the multitude itself, which is weak and misdirected in its passions; hence we "despair," he says, over the "capricious mind of the multitude."[58] Politics here follows by the right of nature, and where there is the immanentism of a naturalized, biological society it is unreasonable women who threaten the fraternity of men and their politics of friendship. One cannot even argue that this is simply how right has formed in particular societies, since for Spinoza this would be the unfolding of an immanent substance (the *civitas*) by force of nature; thus women, by *nature*, Spinoza consistently argues, *must be* weaker, even if this "*must be*," this *necessity*, is born of a logic that itself can admit no empirical counterpoint (hence his dubious claim that *never* have men and women ruled together). To bring what should be a longer discussion to a conclusion, Spinoza's system requires bringing the reproduction of life under the aegis of the political, since this *imperium* first and foremost is the power or right over life and death.[59] Women, as producers of life, must be regulated under the male domination of reason, which in Spinoza is simply the wisdom of reflecting upon the necessity of what is. That is, the bodies of women are *passive* entities that *receive* but do not produce the "general *conatus*," as Negri puts it; they do not have the full life of men; they are never innocent, they never have an *active* conatus, and they are always already guilty of being the object of the *conatus* of men, whose passions are so apparent in the multitude. The feminine, then, is the name for the enemy within the state. They literally embody what will weaken men and the life of the body politic; Spinoza hence likens "consorting with harlots" *and also* "ravishing maidens" by those in power as akin to a "state of war."[60] Thus the state will keep women from rule as a kind of political Viagra for men to regain their vitality, all in the name of keeping life safe; the biopolitical, as we know in the United States, always regulates women in particular in the name of being "pro-life." This patriarchal logic can evince no principle of difference and life must be produced out of "one mind," out of an auto-affection or masturbatory logic that sees the other as a dangerous temptation for the work of this "one body," this "general *conatus*." But such an immanentism would mean political death – states do need to produce the very citizens it is to secure, after all – so it will be up to the state to watch over the reproduction between men and women since, alas, we will never again be "small children [...] infused with an immanent life that is pure power and even bliss."[61] And this vigilance over reproduction and the worries over the weakness of men and their jealous desires is precisely the logic of patriarchal-nationalism, even when written by a "weak, sickly man" forever fascinated by his potency. It is time to leave these forms of naturalism behind, or at least to begin to see how such a life is produced in and through the political, which is always engaged in the "arduous work" of policing the security of life. For Spinoza, no other world is possible, and all that would be political has always already been put in line by nature itself, and politics is to be led, as he never tires of writing, by superstition and by the "sad passions" of hope and fear, since the feminized masses, reduced to pure embodiment, cannot engage in the life of the mind. But beyond this Spinozism life goes on, because finding a space for something other is the work of politics, unworking the immanent systems that have come down to us.

notes

1 Gilles Deleuze, *Spinoza: Practical Philosophy*, trans. Robert Hurley (San Francisco: City Lights, 1988) 12–13.

2 Jean-Luc Nancy, *The Inoperative Community*, ed. Peter Connor (Minnesota: U of Minnesota P, 1991) 12.

3 Michel Foucault, *History of Sexuality, Volume 1: An Introduction*, trans. Robert Hurley (New York: Vintage, 1990) 143.

4 See my *State of Sovereignty: Lessons from the Political Fictions of Modernity* (Albany: State U of New York P, 2012) chapters 3 and 4.

5 Carl Schmitt, *Political Theology: Four Chapters on the Concept of Sovereignty*, trans. George Schwab (Chicago: U of Chicago P, 1985).

6 This is most notable in the work of Giorgio Agamben. See, for example, *Homo Sacer: Sovereign Power and Bare Life*, trans. Daniel Heller-Roazen (Stanford: Stanford UP, 1998) esp. 1–25, as well as the middle sections of *The Kingdom and the Glory: For a Theological Genealogy of Economy and Government*, trans. L. Chiesa and M. Mandarini (Stanford: Stanford UP, 2011).

7 See Carl Schmitt, *The Leviathan in the State Theory of Thomas Hobbes: Meaning and Failure of a Political Symbol*, trans. G. Schwab (Chicago: U of Chicago P, 2008).

8 Spinoza, *The Political Treatise* in *The Complete Works of Spinoza*, trans. Samuel Shirley (New York: Hackett, 2002) 1.7. References to Spinoza's *Political Treatise* denote the chapter and paragraph, respectively.

9 Ibid. 1.1.

10 Roberto Esposito, *The Third Person* (London: Polity, 2012) 18.

11 Ibid.

12 In *The Deleuze Critical Reader*, ed. Paul Patton (London: Blackwell, 1996) 107–13.

13 Deleuze, *Spinoza: Practical Philosophy* 26.

14 Here is the relevant passage from Dickens's *Our Mutual Friend*:

> A disreputable man, a rogue, held in contempt by everyone, is found as he lies dying. Suddenly, those taking care of him manifest an eagerness, respect, even love, for his slightest sign of life. Everybody bustles about to save him, to the point where, in his deepest coma, this wicked man himself senses something soft and sweet penetrating him. But to the degree that he comes back to life, his saviors turn colder, and he becomes once again mean and crude. Between his life and his death, there is a moment that is only that of a life playing with death. (Cited in Gilles Deleuze, "Immanence: A Life" in *Pure Immanence: Essays on a Life*, trans. Anne Boyman (New York: Zone, 2001) 28)

15 Ibid. 30.

16 *Political Treatise* 1.6.

17 This is at the heart of Negri's own Spinozism. See, for example, "Spinoza's Anti-Modernity" in *Subversive Spinoza*, ed. T.S. Murphy (Manchester: Manchester UP, 2004) 79–93.

18 See Gilles Deleuze, *Expressionism in Philosophy: Spinoza*, trans. M. Joughin (New York: Zone, 1990) 173.

19 Gilles Deleuze, *Difference and Repetition*, trans. Paul Patton (New York: Columbia UP, 1994) 40.

20 Spinoza, *Principles Concerning Cartesian Philosophy* in *Complete Works* 197.

21 Spinoza, *Ethics*, trans. S. Shirley (New York: Hackett, 1992) IV, prop. 22. References to the *Ethics* will denote the chapter and the proposition number, along with whether it's located in a proof or scholium, if applicable, as well as the page number for this edition.

22 Ibid.

23 Ibid. prop. 21.

24 Aristotle, *Politics* 1252b29–30.

25 *Ethics* IV, prop. 24.

26 Ibid. prop. 25.

27 Deleuze, "Immanence: A Life" 30.

28 See Spinoza, *Œuvres complètes*, eds. and trans. R. Callois, M. Francès, and R. Misrahi (Paris: Gallimard, 1972).

29 For more on this figure and the connections to Foucault and Arendt, see the introduction to my *The State of Sovereignty* (New York: State U of New York P, 2012).

30 Spinoza, *Political Treatise* 3.2.

31 Hobbes, *Leviathan* (New York: Hackett, 1992) Part II, chapter xix, par. 15, 124.

32 Ibid., Part I, chapter xiv, par. 8, 82.

33 See, for example, *Theological-Political Treatise*, trans. M. Silverstone and J. Israel (Cambridge: Cambridge UP, 2007) 16.17 and 16.18. References to Spinoza's *Theological-Political Treatise* denote the chapter and paragraph, respectively.

34 *Theological-Political Treatise* 17.4.

35 *Political Treatise* 6.6.

36 Even, of course, in the case of the penalty of death, which one can resist even as one has "consented to the law by which they are condemned" (*Leviathan*, Part I, chapter xiv, par. 29, 87).

37 Ibid., Part I, chapter xiii, par. 14, 78.

38 Ibid.; my emphasis.

39 The definition of a multitude below provided by Hobbes can be found in such writers as Dante in his *De Monarchia*.

40 Hobbes, *Leviathan*, Part I, chapter xvi, par. 13, 104.

41 Spinoza, *Theological-Political Treatise* 16.9.

42 *Political Treatise* 16.10.

43 Ibid. 3.5.

44 Ibid. 2.16.

45 Ibid. 5.2.

46 Ibid. 3.2.

47 *Theological-Political Treatise* 3.5.

48 Alexandre Matheron, "The Theoretical Function of Democracy in Spinoza and Hobbes" in *The New Spinoza*, eds. W. Montag and T. Stolze (Minnesota: U of Minnesota P, 2008) 215.

49 Hasana Sharp, "Eve's Perfection: Spinoza on Sexual (In)Equality" forthcoming in *Journal of the History of Philosophy*. My thanks to Sarah Kizuk for providing me with a copy of this.

50 *Theological-Political Treatise* 17.2.

51 Negri, *Subversive Spinoza* 37.

52 Ibid.

53 Ibid. 36.

54 *Political Treatise* 1.5.

55 Negri, *Subversive Spinoza* 28; my emphasis.

56 *Political Treatise* 11.4.

57 See her "Eve's Perfection: Spinoza on Sexual (In)Equality."

58 *Theological-Political Treatise* 17.3.

59 *Political Treatise* 2.17.

60 Ibid. 4.4.

61 Deleuze, "Immanence: A Life" 30.

The sole law of abandonment, like the law of love, is to be without return and without recourse.

Nancy, The Inoperative Community

There is a gift by which one forces the one who receives it to give back a surplus of power or prestige to the one who gives – thus, one never gives. The gift that is abandonment commits the abandoned being to giving without any return in mind, without any calculation and without any safeguard even for his own giving being: thus the exigency of the infinite that resides in the silence of abandonment.

Blanchot, footnote in The Unavowable Community

[...] the *munus* that the *communitas* shares isn't a property or a possession. It isn't having, but on the contrary, is a debt, a pledge, a gift that is to be given, and that therefore will establish a lack. The subjects of community are united by an "obligation," in the sense that we say "I owe *you* something," but not "you owe *me* something.

Esposito, Communitas

kristin hole

THE ETHICS OF COMMUNITY

nancy, blanchot, esposito

introduction

Roberto Esposito's recently translated *Communitas: The Origin and Destiny of Community* extends a conversation about community now spanning several decades. Jean-Luc Nancy, Maurice Blanchot, and Esposito himself all conceive of community in terms of a non-reciprocal obligation that renders the subject ek-static. This self-altering *imperative* in all cases suggests that ethics is a key subtext to our thinking of community. While Esposito's body of writing has centered on questions of the political and in particular on the intersection between the political and life, I am interested in reading for the ethical implications of his writing on community.[1] As his work is consistently taken up in terms of the contributions it makes to discussions of biopolitics or his notion of the impolitical, it comes as no surprise that his entry into the debates on community focuses on the political philosophical history of the concept.[2] Yet given the decidedly ethical timbre of the *munus*, addressing this unexamined aspect of *Communitas* raises important questions about his work that may be otherwise overlooked. The law that the *munus* reveals threatens to subsume difference in its universality, to replace real and embodied others with an abstract duty, and to fix the meaning of community around its impossible obligation.

Similarly, and perhaps more evidently, Nancy and Blanchot can both be read through the concepts of the ethical that underpin their communities. Recent writing on Nancy and community has also focused on the political implications of his thought;[3] I am concerned here with how reading for the ethics of community provides a perspective from which we can parse out key differences, firstly, between Nancy and Blanchot's apparently sympathetic accounts, and secondly, in conversation with the more recent writing of Esposito. As is evident from the epigraphs, the three share considerable overlap in terms of how they conceive community. However, the political genealogy that Esposito traces separates his account from the earlier two. By examining the ethical implications of *Communitas*, I am not arguing that *Communitas* is best understood in terms of ethics, but rather that an understanding of this dimension of the text grants insight into foundational aspects of his argument.

These discussions around community respond to the perceived failure of political communities in the twentieth century and reflect a desire to recuperate community in a non-oppressive way. Thus they are often evaluated as to whether a practicable politics can be extracted from their pages.[4] While these are important questions to pursue, in focusing on the relationship between politics and community, the relationship between ethics and community recedes into the background. My focus on ethics brings into relief differences amongst Blanchot, Nancy, and Esposito that are otherwise less apparent and provides a lens through which to better assess Esposito's contribution. Lastly, bringing ethics back into the discussion of community raises anew the question of the relationship between the ethical and the political. Community provides a conceptual space through which we can begin to address what an ethico-political understanding of our being-here might look like.[5] While community might seem to be a starting place for thinking about a politics that does away with any nation- or identity-based contingencies, it must first be rooted in an ethical sphere that is often seen as a retreat from or at best outside of what we think of as "properly" political.

To fully examine the subtle variations that *Communitas* adds to the thought of community, I address the two texts that inaugurated these discussions and which provide key touchstones for Esposito's thought. Thus, this paper will start by examining the varied conceptions of the ethical behind Blanchot's and Nancy's writing. In many ways these earlier two essays, published in the same year and in dialogue with one another, share considerable overlap. Both turn to lovers as models or limit cases of community. Like Esposito after them, they see community as that which undoes the subject, altering us to reveal the interconnectedness that we are. Yet reading for the differences that the Levinasian inflection of Blanchot's thinking might have for his account in contrast with Nancy's reworking of the Heideggerian *Mitsein* reveals significant conceptual divergences.[6] I focus specifically on difference and transcendence as they operate in these two works because they provide an important link to similar problematics that Esposito is attempting to work through in his account.[7] Instead of focusing on death as a point of comparison between the two, as do many commentators, I examine the figure of the lovers as a site through which we can best read some of the consequences of these ethical discrepancies. This comparative framework will help us to situate Esposito's work and to better evaluate it.

My final turn is to the ethical subtext of Esposito's community, to examine how the notion of duty conditions the concepts of difference and transcendence in *Communitas*. For Nancy and Esposito the death of God means that, contra Blanchot's Levinas, we cannot rely on anything outside of our being-here to ground us or give us meaning. Yet difference must be present within immanence for community to resist totalization or full immanence. Without a beyond to guarantee a form of alterity, Nancy and Esposito will need to find a way of ensuring that difference functions in their respective communities. In

Esposito's account, which in many ways follows Nancy's trajectory through Heidegger, the law becomes central to community. Despite the commonalities amongst *Communitas, The Inoperative Community*, and *The Unavowable Community*, I suggest that Esposito's focus on duty has consequences that give cause for further reflection. The law subsumes difference in its universal obligation and risks missing the exceptionalness of our being-with.

community and alterity

blanchot

Blanchot's *The Unavowable Community* consists of two essays: the titular piece followed by an essay entitled, after Bataille, "The Community of Lovers," focusing on a *récit* by Marguerite Duras.[8] For Blanchot, community persists but not as an entity that can be pinned down in time, space, or language. It is given meaning from a relation to the other that precedes and interrupts the ontological, thus it is unavowable because its reality lies in a sphere beyond representation. Community continues on the hither side of one's own proper consciousness of it, no less real because it cannot be defined or made into a concrete concept. Unrepresentable, this community cannot become some *thing*, an immanence or object that can be used to determine who and what are common or shared, and, consequentially, who or what will be excluded from this sharing. In this way, the community that is unavowable resists totalities and fixed notions of meaning, as both Esposito and Nancy's communities attempt to do. Unlike the latter two, Blanchot's account suggests something transcendent, infinite, and *absolutely* other. Community consists of "an exteriority that thought does not master" (12). This exteriority is transcendent – constituting community while simultaneously rendering it impossible.

The exteriority of community echoes Levinas's ethical relation, which inaugurates subjectivity in what he calls the "face to face."[9] The face-to-face happens outside of any possible experience, somewhere other than and prior to the ontological. It is not an exterior that is in opposition to an interior, but one beyond any simple conceptual opposition, which would remain within the sphere of the ontological. This difference is absolute alterity. The violence of phenomenology, in a Levinasian framework, is that it always reduces the "Other" to the "Same" – the perceiving consciousness of the subject of philosophy. The other is always illuminated for the subject through her own understanding of the world. Even difference is understood relationally. Therefore, the other is always reintegrated into a totality within which alterity is negated in the service of reiterating the Same. To understand the impossibility of understanding, one must see the other with whom I commune as an absolute alterity, transcendent and pulling me towards the infinite. This face-to-face relation is constitutive of the unavowable community.

The other as absolutely so, being outside of the ontological realm creates a problem for representation. Thus the community raises anew, like the ethical relation, the difficult necessity of using words to describe that which lies outside of language and representation. In Blanchot's work this thematizes itself as a concern with language's (in)ability to bear witness to limit cases of experience.[10] In his own writing, Levinas attempts to address this difficulty through the idea of the saying (*le dire*), as opposed to the said (*le dit*), a concept that Blanchot references in his discussion of the transmission of the untransmittable or unavowable of community (18). The said is represented and captured within the ontological, whereas the saying is the ethical encounter, the emergence of subjectivity in the call to responsibility for the other. The representation of the ethical – and community for Blanchot – becomes that which depends on "the saying without the said" or a said that is constantly undone by the saying (*Otherwise than Being* 143).[11] In this formulation, the meaning of bearing witness, or avowing, is emptied out and re-imagined. Levinas writes: "The saying prior to anything said bears witness to glory. This witness is true, but with a truth irreducible to the truth of disclosure, and does not narrate anything that shows itself" (145), and further, "'Here I am' as witness of the Infinite, but a witness that does not thematize what it bears witness of, and whose truth is not the truth of representation, is not evidence" (146). This impossible experience is the unavowable of Blanchot's community, to which he adds connotations of the secret, inspired by Bataille's furtive Acéphale community. By making it something that cannot be captured by representation, Blanchot avoids the problem of reproducing the community as immanent.

In Blanchot's analysis of Duras's *récit*, he explicitly invokes Levinas in conceiving the relationship with the other as a basis of community:

> This is what I read in this *récit* devoid of anecdote where impossible love (no matter its origin) can be translated by an analogy with the first words of an ethics (words Levinas has uncovered for us): an infinite attention to the Other, as for the one whose destitution puts him above all and any being, an urgent and ardent obligation that makes one dependent, takes one "hostage" [...] An ethics is possible only when – with ontology (which always reduces the Other to the Same) taking the backseat – an anterior relation can affirm itself, a relation such that the self is not content with recognizing the Other, with recognizing itself in it, but feels that the Other always puts it into question to the point of being able to respond to it only through a responsibility that cannot limit itself and that exceeds itself without exhausting itself. (*Unavowable Community* 43)

Abandonment to the other and an obligation without recourse or return are here explicitly framed in terms of an ethics. This ethics means that there is no subject, previously constituted, who then gives of herself or relates herself to other subjects. Rather, it is a relation by virtue of which the terms of the relation are brought into being. The relation constitutes the terms it relates, in this way rendering impossible a notion of subjectivity that is first autonomous and immune and *then* comes into contact with others. The being of the subject is constituted relationally as an absolute and infinite responsibility to the other, who permeates and undermines the proper boundaries of the subject, refusing her the illusion of autonomy. Thus Being is only experienced qua being-with. For Blanchot this relation *is* community, "The relationship with the other which is community itself, a community that would be nothing if it did not open the one who exposes himself to it to the infiniteness of alterity, while at the same time deciding its inexorable finitude" (16–17). This infinite obligation that constitutes me, if thought of as that which conditions community, would make impossible community's immanence or the subject's immunization. Nancy is also concerned with reframing community so that it will not be knowable, producible, or possess-able, but his grounding in Heidegger will shape his ethics and his thinking of difference and transcendence in ways at odds with Blanchot's deployment of the terms.

nancy

We have seen how Blanchot's basis in Levinasian ethics relies on both the transcendence of the other and on a concept of difference as absolute alterity. Although his essay is a response to Nancy's *The Inoperative Community*, published earlier in the same year, Nancy's essay takes its titular adjective from a concept in Blanchot's thinking, revealing the fruitful sharing of ideas already at work between the two thinkers. Nancy writes:

> Community necessarily takes place in what Blanchot has called "unworking" (*désoeuvrement*), referring to that which, before or beyond the work, withdraws from the work, and which, no longer having to do either with production or with completion, encounters interruption, fragmentation, suspension. (*Inoperative Community* 31)

Nancy, along with Blanchot, is concerned with refiguring community in such a way that it no longer leads to some property that we would possess or an essence that we would produce in common. It cannot be defined in terms of function or the productive value generated by its members. The community must refuse communion – totality or complete immanence, and be open to constant alteration and interruption. Yet despite similarities in how they approach the question of community, there are subtle differences of inflection between the accounts given by Nancy and Blanchot that I am suggesting relate to the varied relationships each has with more recent philosophical history. Blanchot's Levinasian community and Nancy's reworking of Heidegger can be read through the way in which differing notions of the ethical intertwine at the basis of this community that is not *one*. In contrast with the absolute alterity of Blanchot's other, Nancy's community is one where

> We are alike because each one of us is exposed to the outside that *we* are *for ourselves* [...] I do not rediscover *myself*, nor do I recognize *myself* in the other: I experience the other's alterity, or I experience alterity in the other together with the alteration that "in me" sets my singularity outside me and infinitely delimits it. (33)

Otherness is already present at the origin of the subject, but not because of a relation that would be outside of or prior to the ontological.

In his essay "L'Intrus," Nancy develops these considerations starting from his own experience of undergoing a heart transplant and the subsequent need to suppress his autoimmune system, due to his body's rejection of the foreign heart. These events spur reflection on the ways in which our bodies are always already other to us, from the moment of our birth. It is not the intrusion of a foreign heart that begins the process of self-alteration, but rather the body has always been porous, at risk, and imbued with a strangeness such that it was never properly bounded to begin with. Similarly, I am in community even before my birth and my participation in that community extends beyond my death.[12] I am already other from the moment of coming into the world, and my involvement with others is such that it alters me in the process, revealing that I am singularly plural. Difference finds itself within Being, rather than any beyond.

In *Being Singular Plural* Nancy specifically critiques the notion of absolute alterity in terms of its reinscription of the proper and implication of (theological) transcendence. "It is not a question of an Other (the inevitably 'capitalized Other') *than* the world; it is a question of the alterity or alteration *of* the world" (11).[13] Being does not require interruption from what is properly outside of it, but rather it *is* at any given instant *otherwise*. Furthermore, "all forms of the capitalized 'Other' represent precisely the exalted and overexalted mode of the propriety of what is proper, which persists and consists in the 'somewhere' of a 'nowhere' and the in the 'sometime' of a 'no time,' that is, in the *punctum aeternum* outside the world" (13). Contra Levinas, there is no exteriority, no outside to our being here. This is a different *difference*, based on sharing and division. The French word *partager* carries multiple valences including division and sharing out. To share (out) is to divide. Only by being shared can Being have meaning; otherwise it would be pure presence, a self-enclosed *thing* unable to be *a part* of a shared world. If being is being-with it must involve division, spacing, and giving outward. We can only share because there is space or division between us. Difference thought this way introduces a persistent fragmentation into the ontological. This play of difference is not reducible to a fixed relation between terms – like the relative ontological difference that Levinas worries becomes another iteration of the Same – because the terms themselves are never fixed but rather constantly shared out, spacing and altering in their course.

This community of differences, of singular plurality, cannot be mastered by thought. It refuses any fixed or higher meaning. The continual alteration of Being, its inability to achieve pure presence, inserts something *like* transcendence into the sphere of immanence, a succession of interruptions that prevent Being from becoming a fixed totality.[14] This is in contrast with Blanchot for whom there is an absolute dissymmetry between the (transcendent) other and myself, which is given a mystical inflection through his use of concepts like the infinite. With Nancy, if we can speak of transcendence, it is not in opposition to or outside of immanence, but rather that which prevents us from stabilizing what is constantly shifting in the world. He writes: "Community is transcendence: but 'transcendence,' which no longer has any 'sacred' meaning, signifying precisely a resistance to immanence" (*Inoperable Community* 25). By refusing a transcendence akin to that at work in the concept of absolute alterity, Nancy circumvents the danger of reinscribing the proper that he is trying to undermine. Rather than basing significance in something beyond the ontological, Nancy will argue that we *are* sense and the only possibility for sense. Nancy prefers the term sense to that of meaning, as *sens* connotes that significance is an embodied process, rather than purely cognitive. In this way "sense" connects meaning to the materiality of worldly existence.[15] Sense cannot find its grounding in something outside our world. In French, *sens* also indicates direction and manner, rendering meaning dynamic and relational rather than fixed or closed, possessable or masterable. Our sense is the continual putting into question of sense and in living attentive to this question.

These questions of sense, of humans being those whose being is a fundamental question for them, are the basis of Nancy's "Originary Ethics." I turn now to this later text to examine how Nancy himself articulates his ethical philosophy, in order to better address its relationship to his thinking of community. The essay builds on Heidegger's "Letter on Humanism," exploring the earlier philosopher's claim that fundamental ontology is inseparable from an ethics. As is the case with Nancy's work on community, which pushes Heidegger's description of being as always being-with further along the path it suggests but fails to fully bear out, this essay also edges forward the promising ethical implications of Heidegger's thought. Esposito also draws from the "Letter on Humanism" in one of the few moments in *Communitas* where ethics is explicitly discussed in relation to the idea of community. Nancy builds on Heidegger, arguing that "'Originary ethics' is a more appropriate name for 'fundamental ontology.' Ethics *is* what is fundamental about fundamental ontology" ("Originary Ethics" 189). Ethics may be originary, but it is not a *first* philosophy, after Levinas, in the sense that it is prior to and conditions the ontological. Rather, ethics cannot be thought outside the coming into *Being*: "There is not first a brute fact (the being of beings, the 'there is'), *then* a desire for sense (for this being). If this *were* the case, sense, action, and ethics would have to come after and from somewhere other than the fact of being" (176). For Nancy, then, ethics becomes a matter of what he terms "making sense." The meaning of our being here is precisely undecided and undecidable. "Making sense" is a matter of conducting oneself so as to maintain and heed that responsibility. But conduct as a kind of ethics must be understood in a much more passive form than it may seem to imply. Here it is a question of living with the question, refusing to stabilize its meaning or fix an answer, but maintaining that question in all of its messiness as it impacts the way we are in the world. In fact, Nancy is careful to distinguish conduct (*Handeln*) from production, making ethics, like community, unproductive or *désoeuvrée* (175). Both ethics and community have being-with or our singular plurality at their foundations.

In Nancy's writing, ethics, ontology, and community are virtually inseparable in terms of how they are thought and practiced.[16] Making sense is not a solitary speculative project, but rather "The opening of making-sense is utterly impossible in a solipsistic mode" (181). Here in relation to the ethical, the complex neither/nor of transcendence and immanence is further attended to:

> There is, in principle, neither a simple transcendence nor a simple immanence. If it is entirely legitimate and not simply verbal acrobatics to say that the sense of being is the being of sense, this means that sense (the sense of human existence, but also, and along with it, the sense of the world) is in principle nothing other than action, or conduct. Conduct is thus the proper transcendence of the immanence that is. (Ibid.)

Transcendence as conduct is again something at play within the realm of ontology rather than leading us to a realm beyond our being-here together. The contrasting interpretations of transcendence and alterity between Nancy and Blanchot's communities that I am linking to their differing conceptions of the ethical are best illustrated through how they each take up the shared topos of the lovers.

lovers in/as community

The differences that underpin Nancy's and Blanchot's thinking of community take the clearest form around the figure of the lovers. Despite the shared influence of Bataille, lovers are noticeably absent from Esposito's account of community while having a strong presence in Nancy and Blanchot. As mentioned above, the second essay in Blanchot's monograph is entitled "The Community of Lovers." In this essay, his representatives of community are the (heterosexual) pair depicted in the Duras *récit* who meet in fleeting moments. The woman retains an unknowable and mysterious quality for her partner. This relation between two lovers is one of absolute alterity, always asymmetrical, with the other (woman) signifying the (spiritual) infinity of the ethical space. The two lovers can never possess each other, yet their coming together somehow renders them both *otherwise* than they were before, altered by virtue of their relationship: "[...] it is that strangeness itself, having consideration neither for what they can do nor for what they want, but luring them into a strangeness where they become estranged from themselves, into an intimacy which also estranges them from each other" (43). This language of estrangement is indicative of the furtive and alienated quality of the community evoked by Blanchot's language. His lovers seemingly remain a community of two. This dyadic community reflects its Levinasian influence. Levinasian ethics, while succeeding in framing ethics as a relation of complete responsibility to the other that is incompatible with any form of oppression, is perhaps most limited when we try to imagine ethics in terms of the reality of others with whom we share our world. Many faces populate our daily journeys, and understanding the ethical in terms of this plurality presents more difficulty when using a Levinasian framework. Blanchot's adoption of Levinasian ethical language to think community bears out these consequences, in that the community of lovers carries an isolated quality, taking place between two in nocturnal, ephemeral moments. These meetings are unavowable and thus cannot be captured in representational language. The community of lovers undoes my subjectivity but reads as frustrated and esoteric in Blanchot's writing.

By contrast, Nancy's interest in lovers is in how they act not simply in relation to each other, as a community of two, but as figures traversed by and sharing out towards the larger community. In fact Nancy is critical of Bataille's tendency to reproduce a private/public distinction in his discussion of the lovers. Despite the conceptual work that the lovers might be trying to do, they appear swallowed by their singular isolated passion, existing in a realm inaccessible to the larger community. This conception of the lovers as in some ways in communion or fusion with one another is arguably repeated in Blanchot's invocation and further augmented through Blanchot's use of Levinas. In his own writing, Nancy is worried that Bataille finds in the lovers a *substitute* for community, rather than a limit case or microcosm of a larger relation. Against this tendency, Nancy will argue that

> If lovers harbor a truth of the social relation, it is neither at a distance from nor above society, but rather in that, as lovers, they are exposed in the community. They are not the communion that is refused to or purloined from society; on the contrary, they expose the fact that communication is not communion. (*Inoperative Community* 37)

He further writes of the "sovereignty that lovers, in their passion, expose to the outside rather than produce: they expose it first of all to themselves, to their singular beings, but as singular beings these beings already, as soon as the lovers embrace, compear in and before an entire community" (39).[17]

The lovers bring us back to the importance of difference as sharing, as that which divides out and carries the subject beyond any proper boundaries. The lovers do not complete each other, but what is generated between them moves outwards to the community which thus traverses the pair. They share in their two singularities, but these *two* are already formed by and through community's plurality. The lovers have always already been other because they are constituted within and through multiplicity. They each bring to their coupling a unique complexity, exposing themselves to one another as their singular union is also exposed to the community. Lovers touch and alter each other, but this is only one of many types of touching and processes of alteration that constitute our experiences of being-with.[18] In Nancy's work, the lovers lack any spiritual resonance. Nor is there any connotation of infinity or asymmetry. As with his ethics, Nancy's community is able to push our thinking beyond the dyad, and perhaps for this reason resonates more with our lived realities. Our lives are interrupted regularly by a plurality of others, a changing roster of those with whom we must negotiate space and time on a daily basis. As Nancy writes in *Being Singular Plural*, "This 'other,' this 'lowercase other,' is 'one' among many insofar as they are many; it is *each one*, and it is *each time* one, one *among* them, one among all and one *among* us all" (11). The other must remain an improper noun, in lowercase, or difference would be absolute and the "Other" would hearken to a beyond that for Nancy cannot found or ground us.

The differences highlighted between Nancy and Blanchot through the figure of the lovers also point towards the ways in which Nancy's account provides a much more hospitable conceptual space for thinking about community beyond the human, in such a way that it may include animals, plants, and even non-living elements of our world. The differing ways in which each thinker relates his own work back to Bataille's lovers is reflective of the varied philosophy of ethics underpinning their accounts. For his part, Roberto Esposito creates a community from which the lovers are absent. He shares in and widens the intellectual inheritance of Blanchot and Nancy, including many other voices in his genealogy, and though stemming from a different source, the ethical resurfaces in *Communitas*, to reassert its importance in any thinking of community.

the law in common

We have seen how a Levinasian ethics is central to Blanchot's community, its contours shaped by a transcendent and ethical outside and necessitating that the other be absolutely so. For Blanchot, difference will always be effaced by the Same unless it originates in infinity, ensuring its inability to be captured by our finite representational tools. The inoperative community will also line up with an ethics, but instead informed by Heidegger. As this community cannot rely on any beyond to give it sense, it will locate difference within the ontological. Counter to Blanchot, Nancy insists that dependence on any beyond will always work in the service of reproducing the community as immanent. Transcendence must be found within immanence, in the gaps formed by the entrances and exits of beings for whom change is the only constant. Difference operates as a sharing and dividing out that always occur in the *with* of our existence. Both thinkers gesture towards a community that, like the various ethical systems they rely on, refuses totality and interrupts the self-enclosure of the Western subject.

Alongside Blanchot and Nancy, Roberto Esposito views community as inseparable from the unraveling of the subject. He arrives at this shared sensibility via an alternative route, focusing on community through its political philosophical history. This is not to say that Esposito himself is doing political philosophy. In fact, he shows a consistent concern with distinguishing his own work from the field of political philosophy or any philosophy of history.[19] Despite his forays into Hobbes and Rousseau, the community in *Communitas* is not one that can be encompassed within any traditional notion of the political. Nor does this project translate into a thinking of community as based in ethics. However, the ethical implications of Esposito's argument are evident from the analysis (which is central to his project) of the Latin etymology of the word.

In *Communitas*, Esposito turns to the etymological roots of the word community, to extract the *munus* at the root of our being-with. The *munus* is a gift that is given and not reciprocated; something I give to the other(s) without expecting anything in return. As a giving, the *munus* takes me outside the bounds of my proper subjectivity and leads me through a series of alterations, rendering me otherwise than a coherent and bounded self. Through the *munus* my immunity is undermined and I am put at risk, exposed to and traversed by alterity with all its perilous contagiousness.

Munus contains three meanings, that of *onus*, *officium*, and *donum*. Esposito writes: "In truth, for the first two the meaning of duty is immediately clear: obligation, office, official, position, and post" (4). As for *donum* or gift, it too will take on the semantic valence of duty as "this is the gift that one gives because one *must* give and because one *cannot not* give" (5). This duty is a one-way obligation, a law in the form of a compulsion that forces the subject outside her own proper boundaries. Significantly, the language of the gift and its ethical connotations are confirmed in Nancy, who frames the ethics of making sense in terms of the gift. He writes: "There is no 'fact' before the gift of being, which itself constitutes the gift of or the abandonment to sense" ("Originary Ethics" 178). This is the same abandonment that both Nancy and Blanchot draw from Bataille in the epigraphs, taken from their writing on community. Abandonment, then, becomes the ethical obligation of a giving that takes one outside oneself, without reciprocity or expectation of closure. For Nancy, we are abandoned or given over in being to the task of making sense, whereas for Blanchot the gift is abandonment to the other, referring to the asymmetry of the ethical relation. As we will see, an analysis of *Communitas* reveals that this unidirectional gift is informed by a Kantian ethics rerouted through Heidegger.

Its basis in duty gives the *munus* a Kantian inflection. Duty colors the concepts of obligation and responsibility invoked in defining the *munus*, shaping community around an invisible and difficult-to-articulate law. This brings into relief the fact that the central chapter of *Communitas* is devoted to Kant, and this is effectively the conceptual hinge of the book. After Kant, Esposito will move away from properly political philosophers, taking his readers through Heidegger and Bataille. In this chapter, a tension reveals itself that is notable for my argument, since for Kant (according to Esposito), between the political and the ethical "a gaping abyss opens" (71). Esposito argues that contra many of Kant's commentators, in the latter's writing we must look for an understanding of community not in aesthetics or perpetual peace but in the law itself. He writes: "It's the law and not the will that is at the origin of community, so much so that one could even say that the community is identical to the law" (64). Like Kant, Esposito will ground his community in a form of law or duty, and it is in his discussion of Kant that he will explicitly link the law with the gift.[20] He notes that the law "is the foremost gift of the world as the 'common place' of men: the *Es gibt*, the original giving of itself in which what is diverse appears and exists in relation with that which is other from itself" (65). As evident from the language invoked here, Esposito's interpretation of Kant is heavily filtered through Heidegger.

Esposito reads Kant as the first in his genealogy to fully realize the way in which the law undoes the notion of a self-enclosed subject. The law is prior to the subject, coming from outside and carrying her beyond herself in its traversal. The law reveals that this same subject is constituted by what is other to it – a law that precedes her and which will continue on when her life has expired. He writes:

> The law corrodes; it ensnares us; it breaks down our subjectivity. It comes from outside ourselves, and it carries us outside ourselves both in the sense that one cannot make laws for oneself and, more radically, that the law, which unconditionally commands its nonactualization, prescribes in a certain sense the impoverishment of the subject, with which it is concerned. (75)

In this manner the law inserts a form of transcendence into the world. It is an impossible law in that it can never be possessed, never mastered or fully actualized, or it would lose its transcendent quality. Esposito says of Kant's ethical law that, "Seen from this perspective the categorical imperative isn't only incapable of being accomplished but it is Non-Accomplishment or Non-Actualization itself" (ibid.). The capitalized and thus proper Non-'s reinforce the sense of transcendence that the law introduces into community. Non-actualizability functions itself as a law of the law. Kant's account of the law in relation to community (via Esposito's reading) comes to resemble ever more closely the account of community that Esposito himself will end up with when *Communitas* comes to a close. This makes it particularly interesting that Kant is absent from the appendix, in which all of the other philosophers granted a chapter in the book are present: from Hobbes to Bataille the key points of each interlocutor, Kant excepted, are referenced in the process of elaborating the position that Esposito has developed over the course of the text. Yet the resulting community is one that, after Kant, is constituted around an impossible law; the thing shared by its members in common being the no-thing that links them together, ek-statically.

The centrality of the law alters the nature of alterity in Esposito's account. There will still be an otherness to which the subject is exposed and which prevents the subject's immunization. However, it is not the Levinasian Other or the absolute alterity that Blanchot draws on in his musings on the community of lovers. This would necessitate a transcendence counter to the secular community that Esposito is working towards. Turning to Heidegger after Kant, he further develops his ideas, arguing for "[T]he law's immanence to the subject that it, the law, 'decides.' It is what in *Being and Time* is expressed in the formula 'the calls come *from* me, and yet *over* me,' by which we understand an otherness that cannot be interpreted as either transcendence or as a transcendental" (89). Like Nancy, Esposito wants to account for difference within the realm of Being, opening up a space for something like transcendence within immanence. Esposito will find this space of alteration in the law itself. This situating of difference within ontology follows Heidegger, who critiqued Kant precisely for failing to situate transcendence within Being. In *Phenomenological Interpretation of Kant's Critique of Pure Reason* he writes of

> *the fundamental and crucial deficiency in Kant's posing of the problem of the categories* in general, in *misconstruing the problem of transcendence* – or better said, in failing to see transcendence as an original and essential determination of the ontological constitution of Dasein. Insofar as it factually exists, Dasein is precisely *not* an isolated subject, but a being which is fundamentally outside itself. (213)

Dasein is fundamentally rifted and needs no "Other" to interrupt its continuity.

In his chapter on Heidegger – and I repeat that Esposito, like Nancy, is drawing on the "Letter on Humanism" here – Esposito conceptualizes the law in explicitly ontological terms. He says:

> This means that the community isn't a destination, nor exactly is it a presupposition – archaeology coming together with teleology – if not in the arch-originary form in which the presupposition is a law unto itself. Therefore, it is no longer law in the sense of a *having* to be but in the sense of a *being* that is required. This is the way that the absolute identity of ethics and ontology is to be understood and the reason why it isn't nearly precise enough to argue, as Levinas does, that Heidegger more than once sacrifices the first to the second [...] (*Communitas* 90)

Explicitly rejecting the Levinasian position, Esposito aligns himself with a tradition that draws on fundamental ontology and pushes it further, after Nancy. Situating the law within Being – as a "being that is required" rather than a "having to be" – contains the *munus* within the ontological. Ethics is ontology, but via a Kantian detour. Yet the law must retain a form of transcendence in order to come before and go beyond us, to subject us to its obligatory strictures, and if the law is something that we cannot actualize, own, or achieve. In this way, the law creates an opening in immanence, a play of alteration that cuts through being, such as that articulated by Nancy in terms of singular plurality.

Functioning similarly to Nancy's *partager*, Esposito gets at his concept of difference through the word *condivisione* – a sharing that is also division; a dividing that is always *with* [*con*]. This *condivisione* does not mean that we are divided in such a way that our immunity is enhanced, like the subjects who compose the Leviathan (28).[21] Rather, this dividing-out-with institutes vulnerability and lack in the place of immunized identity. He writes:

> The fact that the Romans understood by the term *munus* only the gift [*dono*] that was made and never the gift received (which was instead denoted in the word *donum*) signals that it is lacking in "remuneration," that its opening cannot be closed by any sort of compensation or reparation if it is to continue in fact to remain shared [*condivisa*]. (139)

This opening out that, according to Esposito, is "always from within and never from without" creates a gap in immanence that institutes a play of difference within Being, which prevents the subject or the community from becoming a totality (139). The *munus* renders us *démuni*, stripped bare from the inside out, lacking, and impoverished.

Yet, despite the similarity between *partager* and *condivisione*, in Esposito's writing, this difference that does not transcend our world is not the singular plural world of others painted by Nancy. Esposito writes:

> The community isn't joined to an addition but to a subtraction of subjectivity, by which I mean that its members are no longer identical with themselves but are constitutively exposed to a propensity that forces them to open their own individual boundaries in order to appear as what is "outside" themselves [...] If the subject of community is no longer the "same," it will by necessity be an "other"; not another subject but a chain of alterations that cannot ever be fixed in a new identity. (138)

Here, the common language of exposure is invoked, but instead of indicating an exposure to the ethical other, or exposure to others within a world of singular plurality, it is exposure to a propensity. This propensity is another way of phrasing the concept of duty, so pivotal to the *munus*. It is the law of community itself – as the law of our being in community – that seemingly renders the subject a chain of alterations. This law, rather than an Other or the many beings who populate the world, dislocates the subject out into community.

Perhaps this subtle shift in emphasis from other beings to the law in Being accounts for the disappearance of the figure of the lovers in Esposito.[22] The law seems to replace the caress, the rub of skin against skin, and the exposure to the mystery of other bodies so present in the previous major texts of the community debates. Giorgio Agamben in *The Coming Community*, another important earlier contribution to this discussion, also makes exposure and love central to the "whatever being" at the heart of his configuration of community. He writes: "The singularity exposed as such is whatever you want, that is, loveable [...] The lover wants the loved one *with all of its predicates*, its being such as it is" (2). Lovers move aside to make way in *Communitas* for the law of (the impossibility of) community. The law is that of our finitude vis-à-vis the law, the impossibility of our doing anything which would *make* community or actualize its inner law.

Making duty the source of our difference leaves the paradox of that which is universal – the law – accounting for our singularity. Can a law that is "in common" fully do justice to the exceptional, the idiosyncratic, and the particular of our being-with? Does it not carry the danger, like all laws, of flattening out specificity and context? In our desire to critique the problematic aspects of grounding community in identity, we must not lose sight of the fact that we enter into a world that is historical and our being-with is always embodied. We are thrown into a place and time in which different kinds of bodies make different kinds of sense. *Communitas* situates community in the relation between beings, or in the dissolution of that relation. It is the between that divides in the sense of *condivisione* and this between *is* community. We are all singular *in relation to* each other, and differences such as ethnicity or gender are produced and effaced in what happens *between* us. The law of community encounters difficulty in addressing real differences and singularities, beyond the abstract notion that we are incomplete and lacking. In this way *division* occludes the very *difference* it attempts to articulate. Esposito focuses on subtractions of identity, but differences tend to multiply.

While influenced by Blanchot's writing – explicitly drawing on his thought in other areas that overlap with that of community, such as the impersonal[23] – Esposito markedly rejects any alignment with Blanchot's good friend Levinas. Yet, despite the self-distancing from the philosopher of ethics, concepts such as the transcendence of the other and the insistence on absolute difference as gesturing towards a realm outside of and beyond Being will remain at play within these discussions of community, through their persistence in Blanchot's important contribution. The unavowable community maintains the desire to move profoundly beyond the self, the search for alterity in mystery rites. In these ways, it keeps alive an important element of Bataille's life and work that might otherwise fall out of view. In Nancy, the mystery – or strangeness, to use a more Nancean term – of the other is kept alive, while grounded in an ethics that needs no beyond to give it sense. With Esposito, the Kantian law altered in its filtering through fundamental ontology points again to something ethical at the basis of the community-that-eludes-totality, but the surprise or mystery that the other is for us is no longer reflected in its pages.

Both Levinas, as a background to Blanchot's work, and Nancy in his writing on ethics make reference to certain laws or maxims in attempting to give a propositional content to their new configurations of ethics. For Levinas, the face of the other speaks and its speech can only be rendered as "Thou Shall Not Kill." The prohibition on killing the other signifies an infinite responsibility to her, from giving her the bread from one's own mouth to substituting one's very life. However, for Levinas, the law as transcendent does not present a systemic problem, as he insists on the transcendence of the other. As long as transcendence is infinite and absolute and not merely understood in conceptual opposition to immanence – thus reducing its difference to the Same – the law can coexist with an ethics of radical alterity. The biggest challenge that the prohibition on killing can produce for Levinas's thinking is that it carries the threat of fixing the ethical into ontological language, and thus betraying its unavowable and unrepresentable nature.

For his part, Nancy discusses the possibility of originary ethics being based on the maxim or law to "respect life" but reassures his reader that this freezing of the content of ethics is in fact a reopening, since we have no idea what life is or what it might mean to respect it. His rejection of any law-based ethics is evident when he writes that

> In principle, the ethics thus announced refers to nothing other than existence. No "value," no "ideal" floating above concrete and everyday existence provides it in advance with a norm and a signification. But this everyday existence finds itself asked to make sense. ("Originary Ethics" 179)

This making-sense is a never finished work, a *désoeuvrement*, which defines us in our being-here-with-others. Ethics as fundamental ontology entails a radical attempt to rethink what ethics might be outside of a rule or law-based system. Esposito re-inserts the law by making it that of our Being. In reopening the meaning of community by returning to language to get the definition right, Esposito may inadvertently fix it within the notion of duty. Unearthing the etymological depth of a word to understand the experience towards which it originally gestured can open up the concept in new ways. But it also carries the danger of freezing its meaning in time. Perhaps this is why we need a community that remains unavowable and *beyond* language, which defines and contains. Relying etymologically on the notion of duty as that which, following his reading of Kant, undermines the subject and renders possible the impossible community, grounds the *Communitas* in a law. The *munus* risks congealing the meaning of community around this impossible obligation.

conclusion

Nancy, Blanchot, and Esposito all render community inseparable from the disruption of the self-sufficient subject. The dismantling of any "proper" subjectivity is the only way towards a community that is necessarily improper, unproductive, unavowable, and un-immunized. I have shown above how this undoing of the subject must be linked to its roots in ethics, be they Levinasian in the case of Blanchot or originary (post-Heideggerian) in the case of Nancy. In *Communitas* the law takes on this function with its ethical connotations after Kant, and works akin to the Other in Levinas, or others for Nancy in terms of their interruption of the subject's plenitude. I have focused on difference and transcendence, to tease out the points of divergence that these varied fidelities produce in accounts of community that overlap and share in many significant ways. For those who, after Heidegger, want to find a way to have difference within ontology, transcendence will always reproduce immanence as totality. For those like Blanchot, compelled by Levinas's critique of the tradition that Heidegger appears to break with but arguably merely continues, we must get beyond Being to make space for that which is truly Other. As Simon Critchley argues, "unless solidarity is underpinned by the separation, distance and radical non-solidarity of the ethical relation to the other [...] then it will ineluctably lead back to an ontological tradition that has shown itself incapable of acknowledging that which resists knowledge, i.e. the source of ethical experience" (252). These positions are irreconcilable, but the tension between them reminds us both that difference must be radical, such that we cannot know or possess the other, and that we must look for meaning *here*, within our worldly existence.

While death tends to be the event around which many analyses of community, ethics, and ontology circulate, it is through the act of loving that we begin to parse out the consequences of these incompatible ethical sources. The lovers are a shared topos in Nancy and Blanchot, through which these nuances are most clearly demonstrated. In turning to Esposito, the notions of duty, obligation, and law shape a community where others recede into the background and the lovers lose their visibility. The focus on duty seems to occlude any sustained engagement with *otherness* as it relates to other beings, how we touch one another, are vulnerable to each other, and are undone by one another. This absence – a consequence of the ethical position that Esposito takes – gives *Communitas* a cold and anonymous feeling, so contrary to the joy and curiosity at work in Nancy's writing. As others fade out, an understanding of difference becomes less relevant, so long as we understand the propensity to which we are subject as rendering us a series of alterations.

The law renders each of us other, but in its universality it risks losing a sense of the exceptional. The surprise that we continue to be for ourselves and each other, and the unexpectedness of love, a new life, a sudden death, or a lost friendship, all reveal our existence *with* in ways that undermine our pretensions to immunity. In that sense it is the exceptionalness of each moment that we touch and expose ourselves to one another that renders the law's mandate impossible. Perhaps we each fail to actualize the law *each* time in a uniquely singular way and this is where an account of difference truly becomes possible within the universal. Allowing that, the law reminds us of the danger of using language to fix a community that must remain elusive and ungrounded if it is to address the historical problems that raise community as a concern in the first place.

While, for some, rethinking community may be the first step towards a new understanding of the role or place of the political, the thinking of community is inseparable from developments in the way that ethics has been framed and refigured since the second half of the twentieth century. It thus raises the question of the relationship (if any) between the ethical and the political anew. It could be that community occupies a privileged site through which we can better come to an understanding of the ethico-political implications of our being-with-here. In its concern with dismantling any stable signifiers of identity, be they group or individual, the community may also be the space where a clear boundary between ethics, ontology, and the political becomes foggy or where these various spheres open up to and alter each other, through a shared exposure that cannot be avowed, will not work, and is highly contagious.

notes

I am indebted to Greg Bird, Jonathan Short, Anne O'Byrne, Dijana Jelaca, and David J. Madden for their insightful comments on earlier drafts of this paper.

1 Although Esposito himself does not conceive of his work as political philosophy, it is undoubtedly a philosophy that is deeply engaged with how the political is thought, be it in terms of an affirmative biopolitics or his concept of the impolitical as a response to militant politics in practice and theory. See, for example, Esposito, "Totalitarianism or Biopolitics?"

2 For examples of this tendency see Barkan; Bosteels; Campbell, "Bios, Immunity, Life" and "'Enough of a Self'"; Campbell and Luisetti; Deutscher; Levinson; Minervini; Neyrat; Oliva. Note that the lack of historical engagement with the idea of community is a fault that Bernasconi finds with Nancy's writing on community, namely that it doesn't offer an "historical account of the metaphysical conception of the community that would necessarily underlie any such deconstruction" (15). Esposito's text can be credited with attempting to provide such an historical account.

3 See, for example, Elliott; Norris. See James, *The Fragmentary Demand* for a nuanced discussion of the relationship between philosophy and the political in Nancy.

4 See again both Elliott and Norris for very recent examples of this tendency.

5 See Gaon who also works in this direction.

6 For other accounts of ethics in Blanchot and Nancy's writing on community see Bernasconi; Gaon. See Ingram for further comparison.

7 See Bird on how Blanchot and Nancy variously approach the problem of immanentism.

8 See Duras.

9 See Levinas, *Totality and Infinity.*

10 See, for example, Blanchot's *The Instant of my Death*; *The Space of Literature*; *The Writing of the Disaster.*

11 On the Saying/Said see also Levinas, "Language and Proximity."

12 See also O'Byrne, "The Politics of Intrusion" and *Natality and Finitude* on Nancy and the body, intrusion, and natality.

13 Although Nancy states in a footnote that this is an allusion to Lacan, the comment applies equally to Levinas in the context of his argument. See Raffoul on Nancy's critique of the Other in Lacan and Levinas.

14 Nancy's concept of "transimmanence" addresses this problematic between transcendence and immanence.

15 See Nancy, *The Sense of the World* and "The Insufficiency of 'Values' and the Necessity of 'Sense.'" See also O'Byrne, *Natality and Finitude*.

16 See also James, "On Interrupted Myth."

17 *Compear* is Nancy's term for the sharing present at the origin of each person. Rather than appearing, we co-appear or compear; the origin is always multiple. See also "La Comparution/The Compearance."

18 See "Touching" in *The Sense of the World.*

19 See Bosteels; Esposito, "Totalitarianism or Biopolitics?"

20 The most relevant texts here are Kant's *Critique of Pure Reason*; *Religion within the Boundaries of Mere Reason*; and *Groundwork of the Metaphysics of Morals.*

21 See Bird, this issue, for a discussion of *condivisione* in Esposito's writing.

22 See also O'Byrne, this issue. The lack of a sustained engagement with sexual difference in *Communitas* is closely linked to the disappearance of the lovers.

23 See Esposito, "For a Philosophy of the Impersonal."

bibliography

Agamben, Giorgio. *The Coming Community*. Trans. Michael Hardt. Minneapolis: U of Minnesota P, 1993. Print.

Barkan, Joshua. "Roberto Esposito's Political Biology and Corporate Forms of Life." *Law, Culture and the Humanities* 8.1 (2012): 84–101. Print.

Bernasconi, Robert. "On Deconstructing Nostalgia for Community within the West: The Debate between Nancy and Blanchot." *Research in Phenomenology* 23.1 (1993): 3–21. Print.

Bird, Gregory. "Community beyond Hypostasis: Nancy Responds to Blanchot." *Angelaki* 13.1 (2008): 3–26. Print.

Bird, Greg. "Roberto Espositós Deontological Communal Contract." *Angelaki* 18.3 (2013): 33–48. Print.

Blanchot, Maurice. *The Instant of my Death*. Stanford: Stanford UP, 2000. Print.

Blanchot, Maurice. *The Space of Literature*. Trans. Ann Smock. Lincoln: U of Nebraska P, 1982. Print.

Blanchot, Maurice. *The Unavowable Community*. Trans. Pierre Joris. Barrytown: Station Hill, 1988. Print.

Blanchot, Maurice. *The Writing of the Disaster*. Trans. Ann Smock. Lincoln: U of Nebraska P, 1995. Print.

Bosteels, Bruno. "Politics, Infrapolitics and the Impolitical: Notes on the Thought of Roberto Esposito and Alberto Moreiras." *New Centennial Review* 10.2 (2010): 205–38. Print.

Campbell, Timothy. "Bios, Immunity, Life: The Thought of Roberto Esposito." *Diacritics* 36.2 (2006): 2–22. Print.

Campbell, Timothy. "'Enough of a Self': Esposito's Impersonal Biopolitics." *Law, Culture and the Humanities* 8.1 (2012): 31–46. Print.

Campbell, Timothy, and Federico Luisetti. "On Contemporary French and Italian Philosophy: An Interview with Roberto Esposito." *Minnesota Review* 75 (2010): 109–18. Print.

Critchley, Simon. *Ethics–Politics–Subjectivity: Essays on Derrida, Levinas, and Contemporary French Thought*. London: Verso, 1999. Print.

Deutscher, Penelope. "Reproductive Politics, Biopolitics and Auto-Immunity: From Foucault to Esposito." *Bioethical Inquiry* 7.2 (2010): 217–26. Print.

Duras, Marguerite. *La Maladie de la mort*. Paris: Minuit, 1982. Print.

Elliott, Brian. "Community and Resistance in Heidegger, Nancy and Agamben." *Philosophy and Social Criticism* 37.3 (2011): 259–69. Print.

Esposito, Roberto. *Communitas: The Origin and Destiny of Community*. Trans. Timothy Campbell. Stanford: Stanford UP, 2010. Print.

Esposito, Roberto. "For a Philosophy of the Impersonal." Trans. Timothy Campbell. *New Centennial Review* 10.2 (2010): 121–34. Print.

Esposito, Roberto. "Totalitarianism or Biopolitics? Concerning a Philosophical Interpretation of the Twentieth Century." *Critical Inquiry* 34.4 (2008): 633–44. Print.

Gaon, Stella. "Communities in Question: Sociality and Solidarity in Nancy and Blanchot." *Journal for Cultural Research* 9.4 (2005): 387–403. Print.

Heidegger, Martin. *Phenomenological Interpretation of Kant's Critique of Pure Reason*. Indianapolis: Indiana UP, 1997. Print.

Ingram, David. "The Retreat of the Political in the Modern Age: Jean-Luc Nancy on Totalitarianism and Community." *Research in Phenomenology* 18.1 (1988): 93–124. Print.

James, Ian. *The Fragmentary Demand: An Introduction to the Philosophy of Jean-Luc Nancy*. Stanford: Stanford UP, 2006. Print.

James, Ian. "On Interrupted Myth." *Journal for Cultural Research* 9.4 (2005): 331–49. Print.

Kant, Immanuel. *Critique of Pure Reason*. Trans. Paul Guyer and Allan Wood. Cambridge: Cambridge UP, 1999. Print.

Kant, Immanuel. *Groundwork of the Metaphysics of Morals*. Trans. Mary Gregor. Cambridge: Cambridge UP, 1997. Print.

Kant, Immanuel. *Religion within the Boundaries of Mere Reason and other Works*. Trans. Allan Wood and George di Giovanni. Cambridge: Cambridge UP, 1999. Print.

Levinas, Emmanuel. "Language and Proximity." *Collected Philosophical Papers*. Trans. Alphonso Lingis. Pittsburgh: Duquesne UP, 1998. 109–26. Print.

Levinas, Emmanuel. *Otherwise than Being: Or beyond Essence*. Trans. Alphonso Lingis. Pittsburgh: Duquesne UP, 1998. Print.

Levinas, Emmanuel. *Totality and Infinity: An Essay on Exteriority*. Trans. Alphonso Lingis. Pittsburgh: Duquesne UP, 1969. Print.

Levinson, Brett. "Biopolitics in Balance: Esposito's Response to Foucault." *New Centennial Review* 10.2 (2010): 239–62. Print.

Minervini, Amanda. "A Note on the Thought of Roberto Esposito." *Diacritics* 39.2 (2009): 95–97. Print.

Nancy, Jean-Luc. *Being Singular Plural*. Trans. Robert Richardson and Anne O'Byrne. Stanford: Stanford UP, 2000. Print.

Nancy, Jean-Luc. "La Comparution/The Compearance: From the Existence of 'Communism' to the Community of 'Existence.'" Trans. Tracy B. Strong. *Political Theory* 20.3 (1992): 371–98. Print.

Nancy, Jean-Luc. *The Inoperative Community*. Trans. Peter Connor. Minneapolis: U of Minnesota P, 1991. Print.

Nancy, Jean-Luc. "The Insufficiency of 'Values' and the Necessity of 'Sense.'" *Journal for Cultural Research* 9.4 (2005): 437–41. Print.

Nancy, Jean-Luc. "L'Intrus." Trans. Susan Hanson. *New Centennial Review* 2.3 (2002): 1–14. Print.

Nancy, Jean-Luc. "Originary Ethics." *A Finite Thinking*. Trans. Simon Sparks. Stanford: Stanford UP, 2003. 172–95. Print.

Nancy, Jean-Luc. *The Sense of the World*. Trans. Jeffrey S. Librett. Minneapolis: U of Minnesota P, 1997. Print.

Neyrat, Frédéric. "The Birth of Immunopolitics." *Parrhesia* 10 (2010): 31–38. Print.

Norris, Andrew. "Jean-Luc Nancy on the Political after Heidegger and Schmitt." *Philosophy and Social Criticism* 37.8 (2011): 899–913. Print.

O'Byrne, Anne. "*Communitas* and the Problem of Women." *Angelaki* 18.3 (2013): 125–38. Print.

O'Byrne, Anne. *Natality and Finitude*. Indianapolis: Indiana UP, 2010. Print.

O'Byrne, Anne. "The Politics of Intrusion." *New Centennial Review* 2.3 (2002): 169–87. Print.

Oliva, Rossella. "From the Immune Community to the Communitarian Immunity: On the Recent Reflections of Roberto Esposito." Trans. Timothy Campbell. *Diacritics* 36.2 (2006): 70–82. Print.

Raffoul, François. "The Logic of the With: On Nancy's *Être singulier pluriel*." *Studies in Practical Philosophy* 1.1 (1999): 36–52. Print.

jean-luc nancy
translated by sarah clift

FRATERNITY

I

The Republic of France is perhaps the only state in the world that has a motto with the word *fraternity* in it. Whether or not it really is the only one doesn't change the fact that this motto continues to be tightly linked to the Revolution of 1789 – after the English and American Revolutions which were more strictly national in character – as the inaugural moment of democracy such as it was being offered to the totality of nations or peoples. As a result, the motto that came to stand for the Republic, not precisely at its inception but doubtless starting from 1793, only really took on its full function and – if one dares say such a thing – its full splendour during the Second Republic in 1848. The historical facts regarding this point are complicated and not very well established, but what is certain is that it took some time for the tripartite motto as it is now known – on the one hand, with the addition of *fraternity* to the other two terms, and on the other, without the supplement of "or death" that was used in 1793 – to be fully adopted. Some groups or people were still proposing other mottos after it was put into place, particularly in the context of labour movements. For instance, the Labour Exchange in the city of Saint Etienne, inaugurated in 1888, bore the motto "Liberty, Equality, Solidarity, Justice."

Because the term *fraternity* is so closely tied to a romantic sensibility (using the term "romantic" here in the broadest sense) and to a way of thinking that goes beyond the strict framework of state laws and institutions, it appeals more to a feeling and to an idea of "community" than it does to principles of social organization. In that sense, one can certainly appreciate the desire to distinguish it from words like "solidarity" and "justice," both of which might be said to develop the implications of the first two terms in the motto, above all, that of "equality."

Fraternity is not looked upon very favourably these days, at least not in France, because it is just too sentimental and has too many familial connotations at a time when the family is no longer a powerful reference. It so happened that when Maurice Blanchot used the word to emphasize the affective dimension of "community," the reference earned him a kind of reprimand (one also addressed to Jean-Luc Nancy) from Jacques Derrida who was voicing concerns about a term that was at once familial, masculine and sentimental, and one that also possessed strong Christian undertones. When all is said and done, nobody – apart from the two just mentioned – seems to have taken it up in the political theory of the last forty years. Quite the contrary, the use of the term some years ago by a candidate in the French presidential elections and then again by the candidate who was subsequently elected (the former president Nicholas Sarkozy) has revived all kinds of suspicions regarding a word that is deemed to be less political than moralistic and more sentimental than sensible.

Topping all of the analyses, we find perhaps the following theme, one which could either direct us against the use of the word or, for some, could be thought to work in its favour: freedom and equality can be understood as the expression of rights but fraternity is not a right. Is it a duty then? Things are not often formulated like that, which gives way to the idea of a hope or aspiration and consequently to the idea of a reality that has virtually no substance, if it is not simply misguided and utopian. It is certainly possible to claim that all the debates surrounding the idea of "utopia" are implied in those concerning "fraternity." One can identify here the enduring trace of an anti-utopian tradition, beginning with Marx who felt that the word concealed an illusion.

II

To raise the question of fraternity in a new way, we must begin by stating two premises: 1) it is unclear whether it is even a good idea to defend this notion, and we must take seriously the suspicions to which its familial, Christian and sentimental characterization gives rise; 2) if there are nonetheless some aspects of the word that are still worth taking seriously, these must aid in a renewed examination of its meaning and, first of all, of the meaning of the family underlying it.

The first premise requires only that we exercise a kind of reserve. One shouldn't take up the idea of fraternity without first considering the possibility that it is not necessarily constrained by the predicates of "familial, Christian, sentimental." Concerning the family, our second premise will lead us to inquire about it. As for Christianity and sentimentality – distinguished from one another but doubtless also implicated in each other – it suffices to say the following: each of those terms names a well-known reality, the one being the dominant non-Muslim religion of the West, and the other being the uncertain, indeed the troubled and perilous, sphere that remains beyond the control of reason.

Though it is not at all difficult to ascribe these characteristics to the ways in which fraternity is understood, they too must be submitted to examination. In effect, it may turn out that they themselves have been determined by certain habits of thought that have become hardened throughout the course of our history.

We shall return to this once we have clarified the notion of "family." First of all, the suspicion that there is something inherently sexist about fraternity arises in the context of the family understood as a patriarchal structure. However, this is not the only structure possible: what we call "family" could be defined as the minimal social group required to generate offspring and to take on what follows from that generation (that is, the care and supervision of children up to the point of their autonomy). Perhaps one could even claim that the way the family reflects or has been brought to the service of powerfully masculine and paternalistic socio-political models has accustomed us to focusing on the father and on transmission to and by males.

Be that as it may, there is something more important at stake: "brothers" are not, first and foremost, those who share the same bloodline. To be precise, "blood" is only the symbol for the kind of filiation that proceeds by way of the transmission of seed (of a natural identity or conformity), which is itself represented according to an ancient schema where the mother had no properly generative agency (but was a simple incubator). "Blood" is far from sufficing as a way of thinking about what generation and filiation are.

Sons and daughters are less those who are united by blood – *pater incertus*, as it was put in Roman law – than they are those who are united through the community of maternal nursing – the *mater certissima*. Whether literal or symbolic, nursing does not consist of the internal, continuous and immediate transmission of a vital principle but in the external, discontinuous and mediated gift of a nutritional substance. Nursing is a process of incorporating foreign substances that the body metabolizes into its own substance. The link with the mother is a paradoxical one whereby incorporation (*certissima*) is the opposite of identification (the infant does not identify with, but *absorbs* the maternal substance into its own autonomous substance); the link with the father is that of identification, not with a body or a substance (*incertus*) but with a figure or a sign.

We have to start from this point in order to reconsider family and fraternity. Brothers – and sisters, but we will come back to this – are autonomous subjects at first, their coexistence being founded on nothing other than the companionship of being nourished (*companion* means "those who share bread") and with no rational grounds for their communal existence. The figure or the sign of the father, which is often called "the law of the father" but would be better called "the father as law," is not determined from the outset. On the contrary: the figure is an empty outline or a sketch, and the sign bears a meaning that is fleeting and indeterminate.

It is certainly possible that the father functions as a completed figure, just as it is possible that the mother does not nourish, or does so badly (all of this is of course to be understood within a symbolic order and moreover "father" and "mother" are not necessarily the genitors or even the legal parents). But this is not the rule. The rule, if we can use this word, would be rather that nothing assures the "common" of the brothers beyond nourishment. The passage to independence that nourishment will have made possible also suggests that we are together by accident, without community of origin or sense (or, to put it in Freudian terms, the "murder of the father" precedes the "father," who is only established as the figure of his proper absence).

III

In this sense, "siblings" [*fratrie*] provide the model for "society" as an association without substantial necessity (ontological or originary). Thus, it is also the model for a "we have to work it out together," rather than a "being together." To find or create an equivalent or substitute for maternal nourishment is a task – or rather, it is a desire – that is both more and less than social: "being" or "sense" is at stake here (and perhaps this passes through art, religion, love, celebration and thought, but not through the socio-political). But to give substance to the figure or sign through which the authority of "the law" shows itself is an urgent and inevitable undertaking, for their initial lack of substance is threatening.

The intention here is not to pursue analyses that would need to be drawn out of these premises in several directions. It is simply to point out the following: "fraternity" does not carry with it *ipso facto* the values of the masculine and the paternal such as we hear them in the ordinary sense. Fraternity expresses coexistence without the necessity of "nature" or "destiny" or "foundation" or "origin." Incidentally this is why the motif of the *enemy brothers* plays a key role in all kinds of mythologies. We ordinarily understand it as pointing to a kind of moral monstrosity, but it actually speaks the simple truth of a relation that is erratic and astray, indeed insane.

At the same time, fraternity also bears the obscure shadow or memory of, as well as the desire for, communal nourishment. It would doubtless be better to refer to "sorority" instead, granting that the fraternal does privilege a masculine one-sidedness. Sorority would be fraternity beyond or below the law, in the sphere or spheres of nourishment – that is, of "eating/rejecting" – that are also the spheres of affect.

Fraternity and sorority intersect, indeed they get intertwined with one other, just as much as they constitute, in a more general way, the masculine and the feminine. Though the bearers of roles are never strictly identical to the complex singularities of persons as well as groups: no one is ever simply or completely a "man" or a "woman," and the notion of siblings [*une fratrie*] does not necessarily signify a manly brotherhood [*une confrèrie virile*]. Perhaps these two terms could be used to distinguish two tendencies in the semantics of "brothers": brotherhood brings together subjects that tend to be identical because they are identified by a function, a trade or a role. Siblings point to the family, one that is, as we have said, only the combination of chance (an encounter) and an embrace (desire) – on the understanding that, on the one hand, the encounter is almost always submitted to preliminary determining factors (social, local, etc.) and on the other hand, that the desire may have been replaced ahead of time, in full or in part, by those determinations. The idea of "marriage" that belongs within the sphere of law (and not to the spirituality or mystics of "nuptials") sums the situation up well: it is a matter of controlling chance or at least – and at the same time – of legitimizing the determinations. One could say that "marriage" is the veritable site and stamp of the birth of the law.

In that case, it appears as if nothing remains of desire and everything has been subsumed by a socio-political organization. But that is only true as a tendency. For one must not forget that the Law – law, the state – is only founded on a retreat of every foundational principle. The figure or the sign of the father, and consequently of fraternity as well, offers an empty space [*la vacance*] that must be filled in one way or another. Brothers are originally orphans who have lost their father, such that nothing allows us to identify them as being associated through whatever it may be – unless it is through the absorption of maternal nourishment, which leads to their emancipation.

From the moment that the paternal vacuum [*vacance paternelle*] – the "power vacuum" [*vacance du pouvoir*] as it is called within the socio-political order – is manifested as such, one must confront the obvious truth that can no longer be concealed by any foundational mythology (a function that is always imperfectly fulfilled, regardless of the mythology). This is the lot of democracy: it must take on this vacuum without appealing to a mythology.

The side or register of the maternal or feminine does not furnish a mythology – at least not for the order of the law, at least not to compensate for the absent father. Desire is not captured by representations. It acts; it enjoys; it sinks or thrusts into the sensible thickness of nourishment: hunger, satiation, and then hunger again, ceaselessly. Or even: life, death. And also: the arts; thoughts; loves, the tremors of being and, if one wants to name them, the gods. It is the lesson that is constantly being taught, from Antigone and Scheherazade to Hester Pryn of *The Scarlet Letter* and then onward to Vera Figner, passing through the *Bacchae* of Euripides.[1]

It is thus not surprising that democracy aspires to open within itself and for itself – for that which, within it, exceeds the strict register of the law – a dimension that itself opens up access to desire or to affect, to what I am naming here in an awkward manner to designate this outside of the law and of power – vacant or not – within which being-together exceeds its proper sociality and its governmentality. If "liberty" and "equality" represent – on the condition of always being rethought – the minimal conditions of a civil association without given foundation, "fraternity" might be pointing to the horizon of this outside of the socio-political. If truth be told, it is not even a horizon: rather, it is an opening in every kind of horizon, in every kind of delimitation. This opening is that of sense: sense in so far as it always returns elsewhere, toward an elsewhere, and never forges a final meaning.

But to remain true to everything that has just been said, I need to recognize that this fraternity would have to be understood as sorority or else as the unbinding of the link between brothers and as the reference that unbinding implies to the law as the fiction of a bond (and as the speech of this fiction) on the one hand and, on the other, to the real of transmission and of the sharing of nourishment, that is, of the affect that ingests and rejects (impulsion/expulsion, impression/expression) the substance of the world. Partaking of this "im/expulsion," the communication of the affect, is still sense (that is, sensible, sensual, sentimental).

Perhaps it is unnecessary to use the term *fraternity* and unnecessary too to use the term *sorority* – for if we were to play out this reversal in too simplistic a way, we would simply be making sisters equal to brothers. But there are not two symmetrical sides: if brothers are truly distinct from sisters, sisters themselves can fraternize with the brothers both in fraternal and in sororal modes. There is no symmetry between the sexes, or there is symmetry only if one considers them solely from the point of view of brothers (in political equality, social equality, etc.).

Fraternity is indeed quite inadequate as a term, even if it isn't necessarily all that dangerous. But it is a signal: it is pointing out that the social, juridical and political order cannot take the register of sense upon itself. It can only open up points of approach or access. But it is essential for it to do that much, and for it to do that much, it must know itself well enough to indicate that beyond the law, sense springs forth.

note

1 Scheherazade opposes her imagination, her mind and her heart to the law of the Sultan; she also acts with the assistance of her sister Dunyazade. Hester of Nathaniel Hawthorne submits to the social law of conjugality, which earns her the stockade and the "scarlet letter." The Russian anarchists, especially the women (Vera Zassoulitch, Olga Loubatovitch, etc.), conceived of their action from the outset as being less political than human – or even "metaphysical" – in the broadest sense (this conforms to the very idea of "anarchism"). Vera Figner writes: "The teaching which promised equality, fraternity, and universal happiness, could not help but dazzle me" (*Memoirs of a Revolutionist* (International Publishers, 1927) 164). In *The Bacchae*, the women of Thebes leave the village for the wild forest upon hearing the announcement of Dionysus' return. It goes without saying that the list could be extended … from Sarah who laughs at God to Simone Weil who, in 1940, knew enough to write that "[e]very new development for the last three centuries has brought men closer to a state of affairs in which absolutely nothing would be recognized in the whole world as possessing a claim to obedience except the authority of the State," or else to those sisters, the daughters of General Hammerstein whose history Hans-Magnus Enzensberger has recounted so superbly.

Sexual difference is one of the major philosophical issues, if not the issue, of our age. According to Heidegger, each age has one issue to think through, and one only. Sexual difference is probably the issue in our time which could be our "salvation" if we thought it through.

Luce Irigaray, An Ethics of Sexual Difference *5*

How are we to break down the wall of the individual while at the same time saving the singular gift that the individual carries?

Esposito, Communitas *19*

anne o'byrne

COMMUNITAS AND THE PROBLEM OF WOMEN

After Cain killed his brother Abel and was expelled by God from the Garden of Eden, the first thing we are told he did was know his wife. She conceived and bore a son named Enoch, after which Cain built a city. For a thinker of the origin and destiny of community who begins his exploration in the Book of Genesis this is the crucial moment even if, all alone, it does not tell a story. Anyone approaching these questions in this text must extend the moment and choose which details to include as essential to the narrative of first founding of a city. For Roberto Esposito, what is essential is the fratricide that preceded the founding and that left its mark on the beginnings of all subsequent communities. This is the starting point of his work *Communitas*. But what becomes of all the other elements of the Genesis story: sex, begetting, bearing and naming? They are left out, and this turns out to be an early indication that sexual difference will also be ignored. This is remarkable. Luce Irigaray was writing as long ago as 1982 that sexual difference was a philosophical issue (perhaps the philosophical issue of the age), but Augustine, whose work is Esposito's point of access to Genesis, acknowledged the same thing in AD 429 and the writers of the biblical text were themselves alive to its importance centuries earlier. This is the single greatest lack in *Communitas*. It is a book that begins at the end of difference.

Of course, difference is everywhere in the work, lodged in every line of the text, but so deeply and quietly at work that one can read the book without being faced with it. *Munus*, gift, donation, the between, the with, self, other, immunity, the proper – all the terms that are vital to the project have meaning only in a world of difference, and yet difference and its sexual origin in the human community never come to the surface. The result is a phallic work that fails to register the anxiety and wisdom of Augustine and the authors of Genesis, who were quite unable to ignore the role of begetting and the pull of desire and, as a result, were unable to quite ignore the problem of women. This is remarkable in a book published in the early twenty-first century, but also – and this will be my concern here – in a book that engages with the history of political philosophy, that is, with a canon where all the great works register the tremor of this anxiety. It is well known that Rousseau – a thinker who had more in common with Augustine than *Confessions* – felt it and, sharing Augustine's wisdom, wrote copiously about women, sex and community. Yet Hobbes felt it too and acknowledged the problem clearly, if briefly. Of the other thinkers treated in detail in Esposito's book, it never comes to the fore for Heidegger (who can hardly be counted among the great political thinkers in any case) but is a clear concern for Kant and wholly unavoidable for Bataille.

However, to pursue this thinking through each of the philosophers featured by Esposito would be to write another *Communitas*, a sister volume that would have to be called *Communitas: This Time with Sex and Women*, or *The Other Communitas*. And this would not be a matter of adding to each chapter an additional excursus, supplementing the existing argument on the origin and destiny of community with argumentation on the origin and destiny of community *in difference*. As Irigaray might warn us, it would be a different work. Indeed, it could be many different works, and it will not be possible to generate one of them here. Instead, I will focus on the *Communitas* Esposito did write, specifically on the passage in the introduction in which he sketches Augustine's contribution to the thought of community, goes on to make a brief reference to Arendt's reading of that contribution, and from there prepares fear as a key element of the modern, Hobbesian thought of community. The passage is short but pivotal. Not only is it the last stop before the modern age, but it is also the moment when the possibility of allowing sex and difference to come to light is foreclosed, and so it is the precise moment when that sister volume does not come to be, the moment of its not being born.

The problem of women is both a distinct problem centered on sexual difference, and a privileged point of access to the problem of difference understood in the broadest terms. Its privilege springs from three distinct considerations. First, the sexual difference that gave rise to each of us – that is, the fact that we each have a father and a mother and each came to be inside a mother's body – suggests that difference is not accidental or in any way supplementary, but lodged in the earliest beginnings of all of us.[1] Second, sexual difference entails desire. It is not that the category of desire is exhausted by sexual desire, and certainly not by heterosexual or reproductive desire, but that no thought of sexual difference can avoid the thought of desire.[2] Third, it is the form in which difference surfaces earliest and most often in the historical canon; it is the form in which political thinkers from Plato to Rousseau and beyond run up against the problem of beings who do and do not belong to the city.

esposito's cain

Esposito's trajectory through the history of political thought and the arc of his argument in *Communitas* begin with Hobbes, in a chapter entitled "Fear." Yet this beginning has been set up in the introductory chapter, where the dialectic of community and propriety that is embedded in modern political thought is displaced by Esposito's ground-breaking, radical excavation of the word *communitas* and the *munus* that anchors it. What he uncovers there is an originary breach or trauma or lacuna out of which we originate as finite and mortal beings (8). He acknowledges that this mortal finitude is no news to the history of philosophy. After all, Plato and Machiavelli both intuited the relation of community and death. He continues:

> Yet it's only with the modern period, let's say at the end of the *res republica christiana*, that

> this fact begins to appear as a problem, indeed as the fundamental problem that political philosophy is obliged to interpret and resolve. Before seeing how it does so, we need to turn for a moment to the Christian conception of community if we are to complete the categorical and semantic frame that functions as the presupposition for the communitarian genealogy under examination here. (9)

If Plato and Machiavelli mark the earliest and latest generations in the pre-modern genealogy, Paul and Augustine represent the many generations of Christian political thinkers in between. (Note that the patterns of inheritance are complicated, with interpretation layered upon interpretation. Already much of Augustine's reading of Scripture is refracted through Paul, and Esposito reaches Augustine through Hannah Arendt.) In Paul's interpretation of the Gospel, the gift or *munus* that Esposito has already identified as central to any conception of community is a gift quite removed from possession and appropriation. According to Paul, we are brothers *in* Christ. We are a community of Christians because we take part in the suffering and death of Christ, and that participation allows us to understand ourselves as taking or appropriating nothing. Although participating in his death allows us to receive the gift of life, it is a gift for which we can never be adequately grateful and which therefore establishes lack at the center of our being.

Augustine deepens the paradoxical relation between gift and lack. Reading Book XV of *The City of God against the Pagans*, Esposito follows his lead to the origins of community in guilt: "the whole world is guilty because of Adam" (*Communitas* 11). How exactly is this so? First, Augustine points out that it was not enough that Adam should have committed the first sin by eating forbidden fruit; sin had to be subsequently fixed in us by the sin of Cain, his firstborn. Original sin is passed down through generations and so our community, as Arendt puts it, is always a matter of "community from and with the dead" (qtd in *Communitas* 11). Second, the origin located in the divine act of creation is doubled by the origin in birth, and original sin is part of that double origin. Creation set man in a timeless union with God, but each of us since Adam is born into history. The second origin (birth) was embedded in the first (creation), Esposito writes, like a thorn or a poisoned gift that proves to be ineradicable.[3] For Augustine, the double origin is essential since it establishes the necessary distance between God's act of creation, on the one hand, and Adam's willfulness and Cain's act of violence on the other. Third, in the aftermath of his violent deed, Cain founded the first city, and, for Esposito, that initial violence would turn out to refer to the violent foundings of all later communities.

Most important for Esposito's argument, however, is Augustine's reflection (in another work) that the Christian love we have for our fellow men is "directly proportional to the memory of common danger (*communis periculi*) that we share" (qtd in *Communitas* 12). This shows the double origin culminating in a doubled fear: of the death that will end this life into which we were born, and from which we can be saved by membership in the community of faith, but also of the other, still more definitive death that would be the loss of eternal life with God. Esposito writes: "Caught in the grip of this double danger, the *communitas* (on the Christian side as well, we should note) acknowledges its constitutive alliance with nothing" (ibid.). Once the response to the nothing is characterized by fear, the bridge to Hobbes is established.

Esposito's interpretation owes much to Arendt's *Love and Saint Augustine*, but the matter of common danger emerges there only after a protracted consideration of the equality that makes the danger *common*. While creation placed *a* man in an alien world, birth places us in a populated, human world. The worldly equality that interests Arendt has its source in the second origin, but it also turns out to have its own double character. As a fact of nature it is implicit in our earthly life insofar as we are interdependent. In Augustine's terms, this interdependence springs from our being alike, which is rooted in our common descent from Adam.

We are all kin and all share the fateful situation of mortality. Yet this equality does not become explicit until we understand mortality as the wage of Adam's sin, so equality comes to mean an equal share in that original sin. Original sin was *both* the origin of our separation from God *and* the beginning of the manmade world which is our home. Arendt writes:

> This sinfulness, conferred with birth, necessarily attaches to everyone. There is no escape from it. It is the same in all people [...] this equality is the predominant fact that wipes away all our distinctions. (101)

Thus society is a fact of nature *and* a product of our historical descent: "The fact that man is by nature a social being means at the same time that, from the perspective of this distinctive human origin, he is familiar with the world by nature and by historical generation" (Arendt 104).

This familiar world comes to be by virtue of our being generated by and into families, and by our going on to generate families in turn. This insight does not require us to specify the shape or character of those families as nuclear or centered on heterosexual relationships or built on the ideal of romantic love; the only specification is that they are the site of our coming to be. That is to say, while our worldly community is certainly community *for and with the dead* (as Esposito emphasizes) it is also community *with those to come*. Arendt acknowledges this in her concern with generation and also, implicitly, in her reflection on the role that ignorance has in the human condition of interdependence. After all, in the absence of knowledge of the other's intentions and inner life, we must rely on belief. She writes:

> This belief in the other is the belief that he will prove himself in our common future [...] The continued existence of humankind does not rest on the proof. Rather, it rests on necessary belief, without which social life would become impossible. (101)

There comes a moment in Arendt's interpretation when the phenomenon of being familiar with the world becomes a matter of the world's *belonging* to the individual (104). The term *individual* is not one she dwells upon – her argument would survive without it – but her insistence on *belonging* stands out in the context of Esposito's sustained argument that appropriation, propriety and the proper are incompatible with community.

While the English words *belonging* and *propriety* have both long been used to connote ownership, they have importantly different etymologies. The language of the proper and propriety – *proprio* in Italian – is derived from the Latin *proprius* meaning what is one's own, special, particular or peculiar. Appropriation – *appropriazione* in Italian – is rooted in the Latin *ad-* (to) and *propriare* (take as one's own).[4] In contrast, the Germanic word *belonging* in its earliest sense means to go along with, to be proper accompaniment to, to be appropriate to. Then the verb *to belong* emerged in English as an intensification of *to long*, thereby incorporating an element of desire and necessary separation. Finally, the archaic adjective *belong* has its roots in equality. The *Oxford English Dictionary* tells us: "[t]he primary notion was apparently 'equally long, corresponding in length,' whence 'running alongside of, parallel to, going along with, accompanying as a property or attribute.'" Thus, the senses of *belonging* and *propriety* are initially quite distinct, and when they do begin to coincide it is in companionship and suitability rather than ownership and appropriation.

In Timothy Campbell's translation of *Communitas*, *belonging* and *to belong* translate forms of *appartenere*, a cognate of the English *to appertain*, which has a different etymology again. It stems from the Latin *ad-* (to, completely), *per-* (through), *tenere* (hold) and so suggests a mode of belonging as a part to a whole, where the part is subsumed by the whole. *Appertain* can also be used to signify belonging to a family, but according to a patriarchal model where the members belong to the head.[5]

In contrast, what Arendt captures in her use of *belonging* is the thought that the world belongs to the worldly individual only because *he* also belongs to *it*; he belongs to it by virtue of being born into it, and thus belongs to the world as to his kin. Thus, when I say "my house" or "my money" I am speaking of property, but when I say "my father" or "my child" I am speaking of those who belong to me *and* to whom I belong. In this way, we belong to a community without its owning us. We are part of it in ways that can be described as appropriate or inappropriate only if these terms refer to our mutual companionship and not to a code of conduct or a norm of comportment. Generational belonging follows this model of belonging with, and, by virtue of our common ancestor Adam, it is how we all belong together, belong to the world, and how it belongs equally to all of us (Arendt 104).

It is not that Arendt ignores death and fear. (Note that she does ignore fratricide, explicitly setting aside Augustine's treatment of the story of Cain and Abel as a complication that would hamper the interpretation she has in view (105 n. 21).) She explicitly takes up the role of fear and danger in the constitution of the human community as a Christian community. She quotes Augustine (who is quoted in turn by Esposito): "For nothing so moves one to pity as the thought of his own danger [...] Peace and love are preserved, therefore, by reflecting in our hearts on the common danger" (Arendt 110).[6] For Christians who long for the world to come, being in the world means being subject to original sin, and also being in constant danger of relapsing into that sinful past. The fear that is shared now has less to do with the end of this earthly life (first death) and more to do with the sinfulness that would turn that death into the loss of salvation (second, eternal death). She writes: "Thus, the necessary coexistence in the *civitas terrena* becomes a free inclining (*inclinare*) toward the other who is equally menaced" (110). Esposito argues that the recognition of common danger is our introduction to the modern, Hobbesian, perception of community (*Communitas* 11).

While Arendt's interpretation of Augustine adds kinship and generation as elements of our social being, and while she turns our attention to familial existence, she manages to do so without mentioning Eve. Perhaps, like Augustine's account of Cain and Abel, references to Eve would have hampered the interpretation she had in view. Or perhaps, for Arendt, Eve was simply beside the point.

augustine's cain, or, the begetting of generations

Augustine is obliged to talk about Eve because he grasps and takes seriously the fact that generations of humankind consist of parents and children, and that without begetting there is no human history. Put another way, his acknowledgment that the human community is community with those who are *not yet* as well as with those who *have been* is more explicit than Arendt's and more central to his approach. Indeed, it shapes the whole task he sets himself in Book XV of *City of God*, the task of relating the history of the two races of humankind

> from the time when the first two human beings began to beget offspring down to the time when the begetting of offspring will cease. For the history of the two Cities of which we are speaking extends throughout the whole of this time or age in which the dying pass away and the newly born take their place. (*City of God* 634)

It is not just the enquiry into the two races of man that signals the importance of difference here, but also Augustine's assuming the responsibility of relating their histories from the initial begetting. Genesis tells us: "Now he (Adam) knew his wife Eve, and she conceived and bore Cain, saying 'I have produced a man with the help of the Lord'. Next she bore his brother, Abel" (Gen. 4.1–2). The bearing of these children required pain and labor because God's punishment of Eve for having tempted Adam into sin was to tie desire to pain: "[I]n pain you shall bring forth children, yet your desire shall be for your husband, and he shall rule over you." This very relation of desire and mastery played out here *between* Adam and Eve is reproduced *within* Cain, their first son. When Cain and Abel presented their sacrifices to God, Cain's was properly offered but not properly divided, which Augustine interprets to mean that Cain offered to God something that he owned, but kept himself for himself (*City of God* 644). Cain was angry at the rejection of his offering, leading God to admonish him:

> "Be still," God said: that is, keep your hand from wickedness, and "let not sin reign in thy mortal body, that you should obey it in the lusts thereof" [...] restrain it with your stillness, "it will return unto thee, and then thou shalt have the mastery of it." (646)[7]

If he is still, that is, if he can refrain from acting in anger, then "the vicious desire of the flesh" that made Cain envy his brother will return to him and come under his control. If he can conquer desire (which he inherited from his mother) using the techniques of mastery (which came from his father) he will be saved for the City of God. But he proves to be his mother's son. God's admonition only made him angrier, his envy increased and he killed his brother. Augustine comments ruefully: "Such was the founder of the earthly city" (ibid.).

Such too are the contingencies of generation. What mysteries of genetics and birth order, biology and time, gestation and birth lie behind the fact that the elder Cain set upon his younger brother, and not the other way around? Why should one have come to make a poorly divided offering while the other gave his all to God? Why should one go on to father Enoch while the other passed from the world, fathering no one? For Cain, contingency continues and proliferates as the number of his descendants grows; for Abel, there is no more uncertainty, because the city of saints endures untroubled by the difference of sex, the desires of bodies, the movement of time. This point is driven home by the arrival of a third child, Seth. Even though Eve conceives him – and conceives *of* him – as a replacement for the murdered Abel, he is identical to neither of his brothers: like Cain, he will go on to be the origin of many human generations and the founder of an earthly community, but, while the community of Cain refers only to itself and is the merely natural aspect of the City of Men, the community generated by Seth points beyond itself to the Heavenly City (637).

It does so by presenting itself as the city without women. In all the generations listed in Genesis from Seth to the Flood, no women are named; they appear only as anonymous wives, daughters, and daughters-in-law. Augustine leads us through the generations and the clues offered by the Hebrew text. *Adam* means *man*, both male and female, and so, "although the woman is called Eve as her own proper name, the name Adam [...] belonged to them both" (669). In the first generation, it is clear that woman is both man and not man. Beginning in the second generation this strangeness is suppressed in the line of Seth, whose first-born is named Enos, which also means *man*, but "its sense is such that it cannot be a woman's name" (ibid.). Thus, this aspect of the City of Man looks forward to the City of God where there will be "neither marriage nor giving in marriage. For there will be no generation there, when regeneration has led us thither" (ibid.). In contrast, three women are named in the line of Cain. The last of them – his last descendant – is Naamah. Augustine writes:

> The line of descent, then, from Adam through the wicked Cain [...] is completed by a woman, because it was by the female sex that the sin by which we all die was initiated. Moreover, what is associated with the commission of sin is fleshy pleasure, which "lusteth against the spirit"; and Lamech's daughter was called Naamah, which means "pleasure." (676)

Indeed, Naamah is the figure who elegantly signifies both the pleasure of sex and the fact "that in the earthly city, even down to its end, there will be carnal generation resulting from the conjunction of male and female" (669).

The origin of all human communities is in the sameness and difference of Adam and Eve. If, on the one hand, we cover that difference by subsuming women under the category human/man, we anticipate the end of community in the stasis of eternal communion with God and the divine mode of being where nothing is lacking. If, on the other, we preserve the problem of beings who belong and do not belong, beings whose belonging *in* the community turns out to be entangled with a proprietorial belonging *to*, and whose equality is thereby compromised by the inequality of mastery and ownership, we remain embedded in the City of Man, that is, in the City of Women and Men.

Esposito reads Genesis (with the help of Paul) and Augustine (with the help of Arendt) in order to discern intimations in pre-modern thinking of the nothingness, lack and fear that will become crucial to his elaboration of the essentially modern combination of immunity and community. The project is successful; perhaps too much so. By the time the first chapter of *Communitas* begins, the Christian fathers have been invoked, the line of inheritance established, and fear has begun to emerge as a central feature of our being together. Yet by directing us to these early texts Esposito gives us access not only to the specific resources he needs but also a great deal more including sex, difference and generation. When the work enters the modern age with the discussion of Hobbes, this pattern emerges again.

hobbes and the right of the mother

In his autobiography, Hobbes writes that his mother was so frightened by the apparently impending invasion of Britain by Spain that she gave birth to twins: himself and fear. In fear we find our most intimate dwelling, according to Esposito (*Communitas* 21). Fear is certainly the element of Hobbes's work on which Esposito dwells; he is not alone in this and the interest is surely warranted by the text. Canetti responds warmly to Hobbes's carrying the thought of power to its extreme, and he appreciates the place granted to fear: "[Hobbes] knows what fear is; his calculation reveals it" (*Communitas* 20). Leo Strauss elucidates Hobbesian fear as a fear of death that is infinitely more powerful than our desire for life (*Communitas* 21). Esposito points to Nietzsche's claim that "'fear is a human being's original and basic feeling; from fear everything can be explained,' including morality which has fear as its mother" (*Communitas* 24). Esposito himself asserts: "The fact is that fear comes first" (21).

First in what sense? This chapter of *Communitas*, leading up to the "Excursus on Hobbes," is a thoroughgoing but purely rhetorical deployment of *primacy* and *origin*.

Fear does not come first historically or ontologically, logically or politically. The argument that Hobbes offers is, rather, that fear is *prior*. Rather than the universal claim that it comes first *of all*, he makes the more deliberate assertion that fear precedes the social contract *in particular*. In his narrative of the emergence from the state of nature, fear is experienced before the contract, and is also a cause of the contract. Yet fear enjoys no absolute priority, and it is not offered as a first cause. Instead, in the context of our everyday tendency to take refuge in hope, Hobbes reminds us that fear comes before hope; in the historical context of political philosophy formed by the thought of Christian brotherly or neighborly love, he asserts that fear is the condition for the possibility of that love; in the context of Aristotelian anthropologies based on the natural sociability of humankind, he insists on the negativity of unsociability. This is what is at stake in the claim in his *De Cive* that "the origin of all great and lasting societies consisted not in the mutual good will men had towards each other, but in the mutual fear they had of each other" (qtd in *Communitas* 24, quoting Hobbes, *Citizen and Man* 113). Yet fear emerges as our condition given a state of equality. Indeed, Hobbes's opening words regarding the state of nature in *Leviathan* make the point clearly:

> Nature hath made men so equall, in the faculties of body, and mind; as that though there bee found one man sometimes manifestly stronger in body or of quicker mind than another; yet when all is reckoned together, the difference between man, and man, is not so considerable, as that one man can thereupon claim to himselfe any

> benefit, to which another may not pretend, as well as he. For as to the strength of the body, the weakest has strength enough to kill the strongest, either by secret machination, or by confederacy with others, that are in the same danger with himself. (183; Book I, chapter 13)

It is a matter of sociability, equality *and* fear. The three are inextricable; each conditions the other and primacy cannot be established among them. These are the conditions of our natural existence. This very existence gives rise to the right of nature, which entitles each to preserve his life, and this in turn gives rise to Hobbes's several laws of nature which eventually lead us to secure our existence under the protection of a sovereign. After all, the conditions of equality and fear join to hinder our sociability, for "men have no pleasure [...] in keeping company, where there is no power able to over-awe them all" (185; Book I, chapter 13).

Here in *Leviathan* it seems clear that the men who share equality and vulnerability in the state of nature are male; they use violence "to make themselves Masters of other mens persons, wives, children, and cattell" (ibid.). Yet the conditions that govern their relationships also turn out to govern relations between men and women, and Hobbes makes this explicit in the context of the question of who can have dominion over offspring: "there is not always the difference of strength or prudence between the man and the woman, as that the right [over children] can be determined without war" (253; Book II, chapter 20). Not only are there women in the state of nature, and not only are they subject to being stolen like cattle and enslaved like men, but they are also to be reckoned with. To avoid the war of a mother and father, the parents must come to an agreement. Thus, not only may women be parties to a contract, but they *must* be parties to the only contract envisioned prior to the social contract. Indeed, in the absence of this contract, dominion over children rests with her. Thus, in the state characterized only by natural right, there emerges in addition a distinct *maternal* right.[8]

At this moment, Hobbes's argument must pause. If the existence of children, women and men in the state of nature gives rise to an unaccountable right, then Hobbes's natural conditions of human existence – fear, sociability and equality – and the right of nature to which they give rise must be re-examined. The problem of women and sexual difference is acknowledged now in the reflection that there are no laws of marriage in the state of nature; if there were, the problem – that is, women – could be enclosed within patriarchal households as would happen in most real existing commonwealths. In the same way, the problems of childhood and generation are acknowledged in the reflection that there are no laws for the raising and education of children, leaving generational difference similarly uncontained. In the face of this, Hobbes reaffirms his state of nature: "[T]here is only the Law of Nature [...]" but immediately goes on to embrace these problems by adding, surprisingly, "[...] and the naturall inclination of the Sexes, one to another, and to their children" (253; Book II, chapter 20).

Now the claim that men take no pleasure in one another's company outside the commonwealth is no longer plausible, since men and women seek out one another for sex and love, sovereign or no sovereign, and parents keep wanting to be with their offspring. The three elements of our Hobbesian natural condition remain in place, but each one has become more complex: sociability must now take into account our partiality to the objects of our sexual desire, whoever they are; equality must be rethought as emerging only when we have survived the inequality and dependency of childhood; and fear, if it is not to be paranoia, must be tempered by the consciousness that we each emerged into adulthood or independence in the bosom of a family, of whatever sort.

Yet Hobbes will not allow his insight to do its work. The structure consisting of contract, first right of nature and dominion remains in place and he understands our first relationships within that structure. It is by the right of nature that I do what will preserve my life, and preservation of my life is the only reason I would allow myself to become the subject of another. Yet each of us came to be in – or out of – a state of subjection. The infant cannot choose to become a subject, but it nonetheless, on Hobbes's scheme of things, owes obedience to the one who chooses to nourish it. The primary right of the mother is tied to the fact that she is the first to see the child and so has the first opportunity to decide whether to nourish or expose it. It is an act of will. She secures dominion over the child by feeding it, but, if she abandons it and it is nourished by someone else instead, that person has dominion. Yet why, in the Hobbesian state of universal hostility, would anyone choose to take on the burden of child-rearing? The most direct response comes in *Philosophical Rudiments Concerning Government and Society* (the English translation of *De Cive*) when Hobbes writes that the condition on which a mother, or another, chooses to raise a child is that "being grown to full age he become not her enemy" (qtd in Pateman 44).[9] Yet it is still an inadequate answer to the question of why raise children at all, and the Hobbesian state of nature remains at odds with the Hobbesian recognition of the mode of human generation and nurture, specifically the need children have for care and our inclination to provide it for them.

In the course of Esposito's treatment of Hobbes, he continues to present the investigation of fear as though it were a question of primacy even though, simultaneously, another strand of thought shows the emphasis of the argument shifting to the question of priority. He sketches a genealogy of fear. Interpreting Hobbes's argument about the emergence from the state of nature – that is, the alienation of all individual sovereignty to the person of the single sovereign – in sacrificial terms, he asks which of the two came first: fear or sacrifice?

> What is this preoriginary origin, this Ur-origin that resides on this side of fear and what in fact determines it? What is that fear born from that is apparently primary but in reality is already second with respect to another event, to another scene, which it is more the cause of than the effect? (*Communitas* 34)

The answer comes from Freud's *Totem and Taboo*, and the pre-origin turns out to be the primal scene where the father is killed by the horde of brothers. This is not a stable origin, and the entire pursuit of the origin of origins is deflected into repetition. The murdered father refuses to stay dead, returning in the form of a guilt that must be borne by all the sons. It is a guilt that can only be expiated by sacrifice, which Freud proposes is initially the sacrifice of the sexual dominance over all the mothers and sisters which the brothers had hoped to gain by their act of parricide. It is surrendered with the institution of the incest taboo. Yet the taboo is not sacrifice enough, and the father must be killed symbolically over and over until the fraternal group is caught in a circle of sacrifice and self-sacrifice where fathers devour sons who devour fathers in turn (*Communitas* 40).

However, the deflection happens too soon, and the circle closes prematurely. The moment of error is unmistakable. Recounting what happens in the aftermath of Freud's original parricide, Esposito writes: "The return of what is most familiar (and who is more familiar than the father) is the same essence of what is 'uncanny'" (*Communitas* 36). The mother, who has oddly but simply not appeared until this point, is now denied the possibility of appearing. Women are mentioned but only as the sex objects dominated by the father and then sacrificed by all brothers jointly. The relation between the brothers and the father that should have been a step in the movement towards origin is understood instead as holding all originary elements within it, and they need only be arranged in the circular patterns of repetition and reciprocity to bring the matter to its close. As Emily Zakin puts it, origination is wrested from the mother (165). In the same gesture, the work of sexual difference is covered over and the brothers appear as though they were offspring of a single parent, as merely their father's sons; rather than plural individuals they are multiple versions of their progenitor and thus a singular horde.

The return of the mother – more familiar than the father, and more deeply hidden – is the eruption of sexual difference and desire. Now the circular movement of sacrificial time is interrupted by a new rectilinear strand. On Esposito's paternal/filial reading, the brothers surrender their own identity in favor of identification with the father who, though murdered, is not quite dead. They give themselves over to a frenzy of cyclical identification that becomes a feeding frenzy in the image of the mythic father devouring his sons, who go on to devour the father from within his bowels. As sons of their father, the brothers are interchangeable; but, as the sons of a father and mother they turn out to be distinct and distinguishable, and their interchangeability is replaced by equality in difference. Thus the double and mutual cannibalism of father eating sons eating father is displaced by the cannibalism and creativity of pregnancy. The masculine sacrificial self-consumption performed in the family-without-women becomes the consumption and novelty that is gestation and birth.

immunitas and *bios*: where's naamah?

At the beginning of this essay I pointed to the moment early in *Communitas* in which its unborn sister volume – *The Other Communitas* – did not come to be. *Communitas* appeared in Italy in 1998 and was followed in 2002 by *Immunitas* and in 2004 by *Bíos: Biopolitics and Philosophy*. Does what I indicate here as not-born in fact see the light of day in one or both of these works? *Immunitas* is a promising possibility. In the book's final chapter, in a detailed consideration of the biological conception of immunity and auto-immunity, Esposito dwells on the specific relation of the fetus to the maternal body or, since there is some slippage in his choice of terms, the relation of the child to the mother. In this relation, the role of difference in the work of generation surges into view: the fetus, whose DNA is contributed by both father and mother, is foreign to the maternal body in which it gestates. Why is it not rejected by her immune system? The conundrum this constitutes for theories of immunity remains unresolved. Or, put another way, the hegemonic biological model of immunity has not been able to transform itself to take this universal phenomenon into account.

The difference that is at the center of Esposito's account here is generational rather than sexual. Certainly human conception does not happen without sexual difference, and he allocates different roles to father and mother insofar as they are understood as two distinct, sexed parental immune systems which contribute in different ways to the mysterious problem of gestation. Yet the descriptions of those contributions are strangely reminiscent of the ancient accounts of human generation that assign activity and creativity to the father and inert nurture to the mother. In this case, it is a matter of an immunological primal scene set in the maternal body, whose contribution to the immune make up of the fetus is regarded as identical to its own; the sperm arrives as the bearer of difference. Esposito elaborates in the context of explaining the fact – unanticipated by dominant immunological models – that for a couple whose immunological markers are too similar, pregnancy will end in miscarriage:

> The antinomic picture this creates is clear: only if the paternal sperm is foreign enough to produce the blocking antibodies will the mother be able to tolerate the foreignness of the fetus – by ignoring it; so much so that to avoid miscarriages, the woman's body is inoculated with the father's antigens. This means that what allows the child to be preserved by the mother is not their "resemblance," but rather their diversity transmitted hereditarily from the father. Only as a stranger can the child become "proper." (*Immunitas* 170)

On the one hand, this is very promising. The thought of the fetus as a semi-allogenic implant in the maternal body, preserved there by paternal difference, yields a sophisticated analogy for those who belong and do not belong to community. It opens up another line of approach to the thought of the proper. Indeed, within a page, Esposito has allowed this model of gestation to break down the distinction between inside and outside, identity and non-identity, self and other, and has located the experience of the stranger at the center of the constitution of the proper. Yet, on the other hand, the fact that the moment of sexual difference is passed over quickly, and without the specific benefits of interrogation from a feminist point of view, means that the new model runs the risk of reproducing patriarchal modes of exclusion by inclusion. In the patriarchal model of politics, if women could not be exiled, they could be contained within private life and kept occupied with the concerns of domesticity; sexual difference was thus acknowledged but assigned its place separate from politics. The same model remains available for generational difference; if children are too familiar to be expelled, their troubling foreignness can at least be contained within the household where their public identity is subsumed beneath the father's name, and their newness and difference risks being molded by a traditionalist education. The stranger may be at the center of the proper, but he may be there in the mode of confinement.

In addition, mothers and fathers, or maternal and paternal bodies, are just part of the story of sexual difference, a part that is quite separable from desire. As we have seen, the authors of Genesis gave both desire and begetting their due, acknowledging Eve's desire and including the woman, Naamah, "whose name means pleasure," in the disastrous line of Cain. Augustine makes careful note of this. Hobbes is explicit about the natural inclination of the sexes to one another, and Freud is clear that the horde of brothers is driven to parricide by desire for the women of the tribe. Insofar as sexual desire is not identical to the desire to beget and conceive (and their separability is an insight embedded in the thinking of all of these authors), it opens towards an understanding of sexual difference as not just a matter of the difference between men's and women's bodies but all the many differences that spur sexual desire.

At the same time, insofar as sexual desire *is* linked to sexual reproduction, it drives the generation of ever more differences across which we relate to one another. Luce Irigaray asserted the central importance of sexual difference,[10] and Elizabeth Grosz has recently developed the thought in these terms:

> Sexual difference is ineliminable, the condition of all other living differences without itself having a fixed identity. Sexual difference is the principle of radical difference, the failure of identity, destination, or finality. It is the eruption of the new, the condition of emergence, evolution, or overcoming [...] If difference is the engine of the world itself – as Derrida and Deleuze seem to argue – then sexual difference is the engine of life, of nature, of all that lives. (*Becoming Undone* 103)

This claim has been controversial, with queer theorists, critical race theorists and others objecting that it relegates queer, racial, ethnic and class differences to second place or a second level of importance. Yet Irigaray's and Grosz's claim is not that sexual difference trumps all others, but that it accompanies all others and is what draws us into a future of more and unpredictable difference.[11] Thus a theory of community that does not encounter sexual difference seems to protect itself from engagement with all real difference.

Yet there is a moment late in *Bíos* when Esposito comes close to just that engagement. The occasion is a discussion of the Nazi suppression of birth. Noting that, biologically, every birth is double, he writes:

> Anything but ordered toward unifying the two (or the many in the one), birth is destined to subdivide the one (the body of the mother) into the two, before the subsequent births in turn multiply those in the plurality of infinite numbers [...] [Birth] doesn't assume or impose but exposes someone (male or female) to the event of existence [...] At the moment at which the umbilical cord is cut and the newborn cleaned of amniotic fluid, he or she is situated in an irreducible difference with respect to all those who have come before. (176)

There is division and plurality here, but once again it is generational rather than sexual, and the role of sexual difference in the process of generation is passed over. Yet, because he has noted that the new arrivals are male and female, the question of how this biological fact might also be a mode of being is already stirring below the surface. Can this difference be understood in ontological as well as biological terms? The question does not emerge on its own terms but is deflected into a consideration of Arendt's thought of natality and its relation to biopolitics. There, the doubling identified with birth is treated as "the doubled point of divergence or noncoincidence of the individual life with respect to that of the species, as well as the single action vis-à-vis the repeated course of daily life," that is, of *bíos* and *zoe* (*Bíos* 177–78).[12]

Even if Esposito does not have an Irigarayan concern with sexual difference, he would recognize her description of the distinctive status of that difference. It is a central claim of Irigaray's that sexual difference is ontologically significant. In *I Love to You* she writes: "Sexual difference is an immediate natural given and it is a real and irreducible component of the universal" (47). In *Bíos*, Esposito takes the same position regarding life:

> History and nature, life and politics cross, propel, and violate each other according to a rhythm that makes one simultaneously the matrix and the provisional outcome of the other [...] Life as such doesn't belong either to the order of nature or to that of history. It cannot be simply ontologized, nor completely historicized, but is inscribed in the moving margin of their intersection and their tension. (31)

This is the structure in his thinking in which Penelope Deutscher, in her article "Reproductive Politics, Biopolitics and Auto-Immunity: From Foucault to Esposito," finds a resource for feminist thinking about political life. If life or sexual difference is strictly a matter of bodies – if, as Deutscher puts it, women's bodies come first – then they can be caught up in the practices and interests of biopolitics; if context comes first, women are forced to choose to act compliantly or actively. Esposito offers the possibility of setting priority aside and letting both life and politics, bodies and laws be generative and generated, simultaneously and together (Deutscher 225).[13]

conclusion

It could be argued, then, that *Immunitas* and *Bíos* are indeed versions of the phantom *The Other Communitas*, but it is not yet clear from these works (1) that the problem of those who are both included and excluded is indeed the problem of women and (2) that the problem of women is recognized as central to political thought. Deutscher's work suggests a way in which Esposito's thought of biopolitics recoiling upon itself is an occasion where the mobility and agility of his thinking opens up feminist-friendly – if not yet feminist – possibilities. One imagined *Other Communitas* could be a broadening and continuation of her work, which, as her essay in this special issue demonstrates, would involve thinking rigorously with Esposito against and beyond Esposito.

Another would be a work that would set the argument of *Communitas* explicitly in the context of the conversation on community that happened among Blanchot, Nancy, Derrida and others in the last decades of the twentieth century.[14] In the case of both Derrida and Nancy, the moment where what I have been calling the problem of women surges to the surface is in their exchanges on fraternity.[15] While both thinkers were and are important influences and interlocutors for Esposito, this conversation and the problems it addresses do not emerge in his own consideration of fraternity in *Bíos* (171–74), which remains focused on brothers and fathers in the terms of *Totem and Taboo*. Derrida, for his part, folds his trouble with Nancy's use of fraternity into the explicitly political turn of his thought, and it is expanded there by the thought of auto-immunity in "Faith and Knowledge" and eventually the work on democracy to come in *Rogues*. In the case of Nancy, the development of this thought might be traced through his 2001 essay-length book, *L'"il y a" du rapport sexuel* [*The "There is" of Sexual Relation*], where the sexual relation becomes the central figure of our being-in-relation.

Yet another would follow the sort of investigation of the relation between biology and politics undertaken by Elizabeth Grosz in works from *The Nick of Time: Politics, Evolution and the Untimely* (2004) to the chapter "Feminism, Materialism and Freedom" in *Becoming Undone* (2011), which I touched upon only briefly here. Yet another would follow the model offered by Carole Pateman's *The Sexual Contract* in 1988, a re-reading of the classic texts of social contract theory with an eye to the sexual contract which must always precede the social version. This work would take the canonical modern philosophers that are the explicit focus of Esposito's book and pose to them – at every turn – the question of women.

notes

My thanks to Greg Bird and Kristin Hole for their very helpful responses to early drafts of this essay.

1 This is not an argument about DNA so much as about how to understand the ontological status of singular beings whose origins are never just singular. See Grosz, *Becoming Undone* chapters 7 and 8.

2 Note that granting privilege to sexual difference does not entail a social or moral program, and also does not trump racial, ethnic, gender and other differences. Again, see ibid.

3 Esposito is not interested in pursuing the lead, but the simile cannot avoid directing us to Eve, that thorn in Adam's side who gives rise to the perennial problem of women.

4 See Bird 287.

5 Ibid.

6 Augustine, *Commentary on Paul's Epistle to the Galatians* 56. The passage is quoted in *Communitas* 11–12.

7 Augustine offers this passage as interpreted by Paul.

8 See McKinney 51–59.

9 Cited by Pateman as *Philosophical Rudiments Concerning Government and Society* in *The English Works of Thomas Hobbes of Malmesbury* chapter 9, 116.

10 It is a central claim of Irigaray's that sexual difference is an – if not *the* – ontologically significant difference. In *I Love to You* she writes: "Sexual difference is an immediate natural given and it is a real and irreducible component of the universal" (47). In *Bíos*, Esposito takes the same position regarding life: "Life as such doesn't belong either to the order of nature or to that of history. It cannot be simply ontologized, nor completely historicized, but is inscribed in the moving margin of their intersection and their tension" (31).

11 See Grosz, *Becoming Undone*, particularly chapters 6 and 7.

12 In Arendt the question of the ontological and biological status of birth and natality also remains below the surface, but only just.

13 However, see Deutscher's "The Membrane and the Diaphragm: Derrida and Esposito on Immunity, Community, and Birth" in the present issue.

14 Ibid.

15 See Campbell, "Translator's Introduction" in *Bíos*. See also Morin n. 7:

> For Derrida's critique of fraternity in Nancy, see *Politiques de l'amitié* (1994: 57 note 1), and paragraphs 4 and 5 of *Voyous* (2003). In this book Derrida also mentions that he had already expressed his concern with Nancy's use of the concept of fraternity at the latter's doctoral defense (2003: 91 n.1). Nancy dismisses Derrida's concern in *Être singulier pluriel* (1996: 44 n.3).

See also Zakin; and Deutscher, "The Membrane and the Diaphragm."

bibliography

Arendt, Hannah. *Love and Saint Augustine*. Ed. Joanna Vecchiarelli Scott and Judith Chelius Stark. Chicago: U of Chicago P, 1996. Print.

Augustine. *City of God against the Pagans*. Ed. R.W. Dyson. Cambridge: Cambridge UP, 1998. Print.

Bird, Greg. "Containing Community." Diss. York U, 2011. Print.

Campbell, Timothy. "Translator's Introduction: *Bíos*, Immunity, Life: The Thought of Roberto Esposito." *Bíos: Biopolitics and Philosophy*. By

Roberto Esposito. Minneapolis: U of Minnesota P, 2008. vii–xlii. Print.

Derrida, Jacques. *Politics of Friendship*. Trans. George Collins. London: Verso, 1997. Print.

Derrida, Jacques. *Rogues*. Trans. Pascale-Anne Brault and Michael Naas. Stanford: Stanford UP, 2005. Print.

Derrida, Jacques, and Gianni Vattimo, eds. *Religion*. Cambridge: Polity, 1998. Print.

Deutscher, Penelope. "Reproductive Politics, Biopolitics and Auto-Immunity: From Foucault to Esposito." *Bioethical Inquiry* 7 (2010): 217–26. Print.

Esposito, Roberto. *Bíos: Biopolitics and Philosophy*. Trans. and intro. Timothy Campbell. Minneapolis: U of Minnesota P, 2008. Print.

Esposito, Roberto. *Communitas*. Trans. Timothy Campbell. Stanford: Stanford UP, 2010. Print.

Esposito, Roberto. *Immunitas*. Trans. Zakiya Hanafi. Cambridge: Polity, 2011. Print.

Grosz, Elizabeth. *Becoming Undone*. Durham, NC: Duke UP, 2011. Print.

Grosz, Elizabeth. *The Nick of Time: Politics, Evolution and the Untimely*. Crows Nest: Allen, 2004. Print.

Hobbes, Thomas. *Citizen and Man*. Ed. Bernard Gert. Indianapolis: Hackett, 1998. Print.

Hobbes, Thomas. *Leviathan*. New York: Penguin, 1961. Print.

Hobbes, Thomas. *Philosophical Rudiments Concerning Government and Society. The English Works of Thomas Hobbes of Malmesbury*. Vol. II. London: Bohn, 1839–42. Electronic. Hathi Trust Digital Library. 15 Oct. 2012.

Irigaray, Luce. *An Ethics of Sexual Difference*. Ithaca, NY: Cornell UP, 1993. Print.

Irigaray, Luce. *I Love to You*. Trans. Alison Martin. London: Routledge, 1996. Print.

McKinney, Audrey. "Hobbes and the State of Nature: Where are the Women?" *South West Philosophical Studies* 15 (1993): 51–59. Print.

Morin, Marie-Eve. "Putting Community under Erasure: Derrida and Nancy on Plurality and Singularities." *Culture Machine* 8 (2006): n. pag. Web. 15 Oct. 2012.

Pateman, Carole. *The Sexual Contract*. Stanford: Stanford UP, 1988. Print.

Zakin, Emily. "Beyond the Sexual Contract: Traversing the Fantasy of Fraternal Alliance." *Between the Psyche and the Social*. Ed. Kelly Oliver and Steve Edwin. Lanham, MD: Rowman, 2002. 159–78. Print.

jonathan short

ON AN OBLIGATORY NOTHING

situating the political in post-metaphysical community

A boundless responsibility alone is equal to a destiny which defies all dimensions of fate.

Jean-Luc Nancy, "Responding for Existence" 2[1]

It is apparent that Roberto Esposito and Jean-Luc Nancy address similar concerns in their work concerning the nature and meaning of post-metaphysical community. Esposito, for his part, takes up the question of the *munus* in terms of the community's *quid*, asking as to the substance of the bond that makes community what it is. At least as Esposito sees it, the question of the *munus* is in apparent contrast to Nancy's inquiries in which the question of relation, of the *cum*, predominates.[2] This articulation, however, seems to raise as many questions as it answers – something that, given Esposito's thinking on the impolitical, can only be deliberate. While such questioning is no doubt highly pertinent to the times and the world we collectively inhabit, I must admit to a certain perplexity that attends defining the bond of community in terms of its constitutive negativity or lack, consigning it to perpetual alteration and self-difference. More to the point, I am perplexed about what follows *for politics* in such a circumstance, and moreover, for a strand of political thought that sees politics as a struggle for emancipation. In this context I wonder if what Esposito calls "alteration" is a purely deconstructive force, something that simply *undoes*, brings to ruin existing political categories without being able to propose new ones. Or is it rather the case that, as one glimpses in Nancy's recent work – work that itself responds to earlier criticisms of political neutrality – a vision of the political as fostering a (post-metaphysical) agency of emancipation also flashes into view? My sense is that these questions are at this point almost impossible to answer, but that the contrast I detect between Esposito's carefully deconstructive reading of Western political categories and Nancy's more affirmative sense of democracy as requiring new types of emancipatory agency might be largely a matter of rhetorical emphasis.[3] Such rhetoric, I suggest in what follows, is in turn a product of Esposito's concern to loosen the grip of what he construes as a tragic narrative in which difference is reduced to the same in the development of the legal person, finally becoming synonymous with the hypertrophy of community itself. My argument will be that Esposito's narrative tends to eclipse the compositional difference that is repressed, but which in the end must return to demand its share. This share is not a share in a pre-determined communal arrangement, but functions according to what Nancy will call the law of the in-common as the exigency of justice.

To unpack this argument will require several stages. In a first section, I detail the relationship that Esposito establishes between an original sense of community as obligation and sharing (so quite close to Nancy's reading of community as shared being), and the modern communitarian understanding of community as an essence that is shared in common. These two different notions of community are established through the concept of the proper, which serves as their pivot-point or hinge, and which for Esposito develops through the legal category of the person as it is appears in the Roman world onward. A second section contends that Esposito's reading of the person and the fate or destiny of community emerges through his reading of Heidegger, first as a deconstructionist of the metaphysics of community, and second as going against Heidegger's own insights as expressed through his fatal political allegiance. This engagement is largely responsible for Esposito's attempt to rhetorically situate the political within community in negative and impolitical terms, and hence his apparent reluctance to articulate being-with in the more positive and liberating registers that Nancy does. In a third section, I return to Esposito's category of the person as a figure of the (tragic) communal-political history of the West as the unfolding of immunity, contrasting it with Giorgio Agamben's parallel discussion of the *persona* as inaugurating a "comedic" history in which the singularity of the individual is able to turn against the legal person in a liberating gesture. Finally, in a fourth section, I enter into a brief but direct comparison with Nancy's recent writing on the relationship between communal being-with and the political, arguing that Nancy shares with Agamben a "comedic" and emancipatory view of communal and political exigency that offsets Esposito's own tragic and correspondingly somber rhetorical construal of the need to displace the political via its impolitical double.

I

According to Roberto Esposito, the familiar concept of community as a property or identity held in common by its members represents a significant departure from the term's original sense. At the beginning of *Communitas* Esposito outlines this original sense by way of a conceptual retrieval of community's root "*munus*" as it appears in the Latin of ancient Rome, although he also relates it to the older Greek term *koinos*. In these instances, the original sense of community, which Esposito will distinguish sharply from its modern form, is that of expropriation. Neither a common substance, nor an essence or thing, the *munus* has the sense of constitutive exposure to and expropriation by otherness. Here, to belong to a community is not to bathe in an essence but to be exposed on the line of relation to a foreignness or an outside that is already internalized; to share in common means not to belong just to oneself but to be drawn outside of oneself toward that relation to otherness in which one shares. To share in is thus also to share out, to participate intrinsically to risk expropriation. Thus it is impossible to think of community in this sense as allowing for the stabilization of an identity that would be proper or essential. Rather, as Esposito's discussion of the Latin terms related by use to the *munus* show, i.e., those of *onus*, *officium*, and *donus*, the stability of one's identity is secondary to a fundamental obligation to (respectively) be responsible for something, to be in service to another, and to give something (back) to others. According to Esposito, this retrieval of conceptual import shows that the ancient *communitas* was marked not by fullness and self-sufficiency but by lack. As he writes:

> *communitas* is the totality of persons united not by a "property" but precisely by an obligation or a debt; not by an "addition" but by a "subtraction": by a lack, a limit that is configured as an onus, or even as a defective modality.[4]

So much is this the case that it is only with the inversion of *communitas* that we might find the transcending but reciprocal condition of *immunitas*. The one who is immune is specifically exempted from the expropriating obligation imposed by his or her community, and is thus outside the negativity and lack constituting it, free to "preserve his own position," that is, to foster an apparently independent identity outside the community's expropriating webs of obligation.[5] As this certainly implies, the condition of *immunitas* goes along with a position of social and political privilege, a feature that will be taken up in the subsequent discussion of personhood.

In his more recent work Esposito has treated the concept of the person as both the bridge between the edifice of Roman law and modern notions of legal rights, as well as the point of transition between the ancient meaning of community and the concept of immunity that comprises the community's modern form. Together these points of contact suggest that the concept of the person is central to the constitution of the modern community insofar as it is the person which forms a real unity between archaic and modern forms of life, however much particular aspects of such forms may have changed or shifted with time.

The concept of the person in ancient Rome, for Esposito, is the original site for establishing the social and political propriety of the individual. The individual who is "proper" is qualified to exercise authority and domination over others, to possess property, and to be exempt or immune (at least provisionally) from the expropriation that the ancient community imposes on its members. From Rome onward, the legal category of the person is fundamentally comprised of a mode of domination which turns out to be the nascent expression of the modern immunitary social order. Esposito traces a brief but telling history in this regard. In Roman law, the notion of the person consists of two elements that are superimposed on one another: "a biological body and a site of legal imputation, the first being subjected to the discretionary control of the second."[6] Because of this split it cannot be said that the person is simply the same as the human being, but rather that the person is a social and political substance in which a biological, bodily element is subordinated to what is considered "the rational and volitional, or moral part of the individual, the part invested with a universal value."[7] The function of this rational part is to dominate and bring under its control the non-rational or biological features of the human being; only those who can do so successfully can *properly* be considered persons, those accorded the full protections and normative entitlements of socio-legal order. Thus, as Roman law would have it, no individual is a person by nature, but rather must qualify as such. In the ancient world, the "lower" animal (or especially the vegetative) parts of the soul are opposed to those of the "higher" rational part, and this is what also constitutes a political model of justice, that is, the proper ordering of social relations based on the domination of the many by the few.

If this is so, then in the Roman world the frontier between natural human life and the person is a porous and mobile one, the key purpose of which is to justify the subordination of everyone who does not qualify as a legal person. Esposito outlines this in two ways. First, he says, it is evident that in Roman law a complex continuum exists between persons and various kinds of non-persons. The slave, for instance, is permanently divided between certain aspects of personhood, such as the capacity to manage estates or carry out functions requiring a high degree of skill. Yet it remains the case that the slave can never be a person, since slaves are by definition

> eternally suspended between the condition of person and that of thing. The slave was a thing with the role of a person and a person reduced to the status of a thing, depending on whether we look at the actual functions performed in Roman society or at the strictly legal classification.[8]

Thus, while the slave is legally able to testify in court, and whose death at the hands of anyone other than the owner could be considered murder (at the least entitling the latter to compensation), the same slave could be dispatched at will by their owner. For the slave-owner, the slave is a mere piece of property, not a person, while for the rest of society, precisely because of this status, the slave takes on effects of personhood by virtue of the owner's status as a person. One can see here quite clearly the link between the valorized element of personhood and its connection to the proper (and so to property), on the one side, and the disqualification and subordination of the improper (the slave as thing and thus property), by those constituted as proper, on the other. Esposito's second example, within the continuum between the proper and improper, illustrates that the person, and hence the right to the status of the proper, is never something anyone simply has by birth. To substantiate this claim Esposito invokes the infamous right of the *Pater* to put to death any of his children, but particularly his male heirs, precisely those ostensibly destined for legal personhood. According to Esposito, if no one is by nature a person, this is especially true of the son, who

> before he became *pater*, that is, a legal subject, he still had to pass through the status of *filius in potestate*, which reflected the fact that, in the changeable dispositif of the person, since human beings arrived into life from the world of things, they could always be thrust back into it.[9]

Because the threshold between the person and the human being is essentially mobile and shifting, functioning as a political technology of qualification and disqualification, no one can ever be a person unconditionally.[10] But this is equally true of persons themselves, who by virtue of disgrace or incompetence could be stripped of their status and so forfeit access to the proper, becoming once again mere things. Even in the highly stratified Roman world of persons and non-persons, the original sense of the *munus* as the expropriating power of the community shows through. If no one is a person unconditionally, then no one is ever entirely immune to the expropriating logic of communal obligation.

Beginning with Christianity and extending into modernity, however, everyone is supposed to be accorded the dignity of the person, with the consequence that everyone (at least formally) qualifies for unconditional immunity against the communal obligation once (conditionally) reserved only for the few. Yet as Esposito points out, despite the differences between the social orders of the ancient world and those instituted by Christianity and after, the conceptual-legal dispositif of the person remains unaltered in its basic structure, internally dividing the human being into superior (proper) and inferior (improper) parts. The Christian theological-philosophical doctrine of the human being as an entity split between an immortal soul and a physical body repeats and reinforces the legal category of the person that the medieval world inherits from Roman law. On the basis of the latter it is possible to constitute new theological-political hierarchies and modes of attendant disqualification from the proper on the basis of one's proximity or distance from divine grace and what it enables, the alleged mastery over corporeal appetites.

Although for its part the modern world reworks the concept of the person, supposedly making it an inclusive concept extending human rights to all, the former does not question in any substantial way the split that internally constitutes the human being as person and which allows the mechanisms of qualification and disqualification to continue to function. These mechanisms constitute the community as a substance distributed out in unequal shares on the basis of one's proximity to or distance from the proper, which in turn is defined according to the category of personhood. Thus the system forms a self-reinforcing circle, with the proper founded on personhood and personhood buttressing the proper. One of the ways in which this situation plays out is that in modern liberal theory – and here Esposito echoes Hannah Arendt's critique of human rights – the person who is accorded imprescriptible rights presupposes as its inverse the disqualified bodies of those who are merely alive.[11] This situation obtains because, as Esposito argues, at least since Mill (although one could add since Locke), liberals have conceived of human life as composed of "a bodily thing or a reified body" over which "the person exercises his or her proprietary dominion."[12] The person of liberal legal theory is composed of a de-individuated, thing-like corporeal substance, and an individuated rational element that is qualified to demonstrate the property of mastery over its body and mastery over this body conceived as its property. What this liberal category of the person ultimately sanctions, just as it did in ancient Rome, is the ontological valorization of some lives over others. In modernity, liberal politics makes it impossible that human rights can be provided universally through the legal-political category of the person, since it is the latter that already serves as a device of disqualification. There can be no universal personhood, in other words, because the category itself presupposes a zone of non-personhood or disqualification. A device of exercising sovereignty over human life, for establishing and preserving the proper, and thus for providing a key part of the modern apparatus of immunity from collective obligation, the dispositif of the person is intrinsic to what Esposito calls an "immunitary current aimed at the preservation of a privilege threatened by those who are excluded from it."[13] The dispositif of the person is by its very nature a political technology that irreducibly splits human life into proper and improper elements, and on this basis, within the community, assigns to the proper part "an ontologically superior status" allowing it to disqualify and subordinate those elements deemed improper.[14]

It is at this point that we can link the dispositif of the person to what seems to be its apparent opposite, the modern community essentially conceived. In contrast to the ancient community of expropriation, the modern conception of community consists in what Esposito calls the "affirmative entification" of the original communal relation by means of its substantialization, imputing to it an essence or common property.[15] The contemporary notion of the communal relation is described by Esposito as "mythic drift," by which he means that the constitution of the community as a shared essential property is but an inverted amplification of the immune relation embedded in personhood itself. That is to say, if the legal individual secures a relation of immunity with respect to the expropriating community by "entifying" the individual into the person, the modern version of community does the same by conceiving community as a meta-person, a sum of individuals united in the propriety of an essential trait through which "they are the owners of what is common to them all."[16] As Esposito argues, "these conceptions are united by the ignored assumption that community is a 'property' belonging to subjects that join them together [...] conceived of as a quality that is added to their nature as subjects, making them also subjects of community."[17] Once community is conceived in terms of a common property it cannot help but become, like the concept of the person, both exclusive and divisible. Consequently, according to Esposito, community becomes mired in the dialectic of a mythical "lost and found" with respect to the proper it seeks to instantiate but which turns out to be always out of reach. As Esposito points out, "It is the dialectic of lost and found, of alienation and reappropriation, of flight and return that joins all *philosophies* of community in a mythology of origin" that disposes them to violence.[18]

On the line of investigation pursued here, one can find in Esposito's discussion of personhood a narrative in which the original sense of community in the ancient world, the *munus* of obligation, turns into its opposite, namely, into the community founded on an essential property through the extending of personhood to all members of the community qualifying as proper members. The transition from the ancient community of obligation to the modern community of essential property occurs through the hypertrophy of the category of the person, and through it the drive to immunity. If immunity was once limited by the power of obligation, modernity creates community in the form of radical autonomy from, and even hostility to, otherness. The modern form of community cannot help but be violently exclusive as long as it is implicitly modeled on the concept of the person, the very form of exclusivity and immunity. This proximity between person and community even suggests to Esposito a parallel relation between liberalism and Nazism. Esposito suggests that the liberal conception of the person is conceptually resonant with the Nazi conception of the biological community, even though Nazi ideology rejects the rational-spiritual connotations of personhood by reducing the human being to a biological substrate; the latter move, Esposito argues, simply substitutes a communal biological substance for the displaced legal-spiritualist element. Notwithstanding this substitution, the liberal person and the Nazi biological substrate are mirror-images of one another. In deciding on the qualification or disqualification of life, of membership in the community, Nazis and liberals share more than what separates them through their common inheritance of the category of the person and its immune functions. As Esposito tellingly comments, despite the undeniable differences between liberalism and Nazism,

> this does not eliminate an unconscious semantic affinity between them: it is rooted in a conceptual vector that has been around for a very long while and is therefore resistant to the shocks and reversals it has undergone over time. Observed from this standpoint, where even opposites find an original point of tangency, the formal abstraction of Roman law seems to topple over into the concrete immediacy of biopolitical power.[19]

II

Although it is not readily apparent from the preceding discussion, Esposito's exposition of the category of the person and its role in shifting community away from obligation and toward immunity is intimately bound to an engagement with Heidegger's thought. An important gesture in this direction is supplied by Esposito's contention that it is neither possible nor desirable to return to the ancient notion of community we have apparently lost. Such a gesture of nostalgia would be of the most conservative kind, and would inevitably draw Esposito's work on community toward the sort of essential community that is the object of his critique. In a significant way, that he does not resort to a gesture of this kind follows from his reading of Heidegger as a radical critic of the metaphysics of the proper, and thus, first of all, as a critic of exactly the kind of conservative gesture of return that Esposito associates with contemporary political philosophies of community. Taking his leave from these latter, Esposito writes that "Heidegger's isn't a political philosophy but, more specifically the deconstruction of a political philosophy in the thought of community."[20]

Heidegger's thought of community, as Esposito conveys it, radically exceeds all thought of association – including all forms of political association – beginning with already-constituted subjects, identities, or persons, who then subsequently relate in some way through their common share in the proper. This traditional notion of relation is what unites the otherwise apparently different conceptions of the communal relation as based either in the shared empathy of custom or rooted in commonly held standards of rationality. According to Esposito, Heidegger's innovation in this regard is to conceive of the relation between subjects as presupposing their being-in-the-world (*in-der-Welt-sein*) which thoroughly establishes beforehand the very possibility of any subsequent relation. If relation between subjects is possible, in other words, it is possible only on the basis of a world shared not through those subjects but as their prior condition, as constituting the very domain of togetherness in which something like a relation could occur.[21] For precisely this reason, Esposito argues, community, thought in terms of this ontological domain of prior relatedness, can only be thought as a simple "with," as *mitsein*. The existence of each member of the community, Esposito comments, "cannot be declined except in the first person plural: we are."[22] The mistake of what we previously called modern community is to have tried to add the "together" after the fact, whereas "The *cum* isn't something that is added on from without to the being of existence. It is that which makes being the being that it is."[23]

This reading of Heidegger structures Esposito's explication of his ontological view of community in the following terms: community is what can never be construed as thing-like, as an object of knowledge, precisely because it is only a relation. Yet as a relation that precedes and makes possible every relation between existing entities (the *relata*), community can only appear as *nothing*, meaning that it is "essentially unfinished, that its incompleteness is its essence."[24] The community, to speak Heideggerian, is essentially ecstatic, always exceeding or transcending itself, and so is never given in the field of present things. This in turn means that the in-common of community, what is shared, is also what no one possesses, and what can never be possessed. This entails for Esposito that the sharing of community can only be conceived in terms of lack, never as fullness, and this lends to it a decidedly negative character that cannot be articulated in positive terms: "We are not joined by a fullness but rather by an emptiness, by a defect, by a fall."[25] It is this intolerable emptiness or groundlessness that Esposito believes the metaphysics of the person, and thus the immunity of the proper, develops in order to conceal.

For Esposito it is this negativity or lack that mandates that we think community in an exclusively subtractive or deconstructive sense, as the disruption of all attempts to realize community as reified actuality or embodied in some representative scheme, in other words, to make it coincide with an individual or collective subject. But as Esposito argues, the only properly political possibility of this sense of community is impolitical, since we can only "appropriate for ourselves the origin in the form of its negative, of what community is *not*."[26] As Esposito goes on to explain, the impolitical here has the following sense:

> The community is and needs to remain constitutively impolitical in the sense that we can correspond to our being in common only to the degree in which we keep it away from every demand for historical-empirical actualization, that is, if we do *not* take on for ourselves the roles of subjects: the community cannot have "subjects" because it is the community itself that constitutes – that deconstructs – subjectivity in the form of its alteration.[27]

This claim is the entry-point for the impolitical as that negative moment arising within the political but remaining in excess of it, which blocks or interrupts the immunitary drive to historical-empirical actualization. There is thus a close parallel between Esposito's category of the impolitical and his category of the impersonal. In this sense, the impersonal – again not unlike his concept of "affirmative biopolitics" introduced at the end of *Bíos* – is not a frontal assault on the concept of the person, but rather a kind of sideways subversion through the location of "alteration" within the proper. As Esposito writes in this regard, the impersonal is not to be conceived as "simply the opposite of the person – its direct negation – but something that, being *of* the person or *in* the person, stops the immune mechanism that introduces the 'I' into the simultaneously inclusive and exclusive circle of the 'we.'"[28] Elaborating on this idea, Esposito claims that

> The impersonal is a shifting border: that critical margin, one might say, that separates the semantics of the person from its natural effect of separation; that blocks its reifying outcome [...] the impersonal is its [the person's] alteration, or its extroversion into an exteriority that calls it into question and overturns its prevailing meaning.[29]

The impolitical, like its correlate of the impersonal, calls into question the propriety of the proper in the person and the community by subtracting or blocking the element of proprietary identification.

No doubt, as Esposito argues further along in this section of *Communitas*, his impolitical repudiation of the realization or actualization of community is directed against Heidegger himself, especially against the latter's reactionary attempts to "realize" community during his tenure as Rector under the National Socialists. As Esposito comments, "It is as if Heidegger misunderstood himself, as if he lost the most vital dimension of his own philosophy" in siding with the Nazi attempt to establish the properly German community within the university and the nation.[30] Esposito apparently seeks to use Heidegger against Heidegger, and by extension to establish a radical vigilance against any metaphysical political project of establishing the proper on the site of the lack that is community. Heidegger's political misadventures furnish an important lesson that Esposito has no doubt taken to heart: every attempt to think the meaning or sense of the community in a positive form can only reintroduce the metaphysics of the proper, to reduce community to a presentable and actual thing. But this justification for permanently interdicting the metaphysical community situates the impolitical precisely within the zone where the posture of political vigilance and political quietism apparently coincide. This is because the impolitical sense of the political is just the attempt – as Esposito's comments above indicate – to stave off the worst, to forestall the solidification of the proper in the person and in the community.[31]

What Esposito presents with particular clarity here is the issue of ontological neutrality or indifference that appears to make untenable for ontology the affirmation of any political project. The ontological ground of community as relation, and hence as what can only appear in the standard "ontic" domain of politics as negative, as an inversion, can never appear as such without becoming immediately subordinated to that domain as one of its instances. The ontological in Esposito's presentation can only remain such by withdrawing, remaining aloof and indifferent to any and all of its products. This in turn seems to suggest that – at least on Esposito's take on Heidegger – any position within the political domain is indifferently enfolded within the ontological relation, and, if so, it is impossible from an ontological perspective to espouse any political position as being superior to any other: the ontological cannot be the domain of value and evaluation, only that of coming to pass in relation.[32] Put this way, it is unclear what kind of affirmation Esposito's affirmative biopolitics – and for that matter his strategies of the impersonal and impolitical – might yield, aside from an unstructured, unlimited affirmation of an ontological "drift" that complicates but never poses an alternative to existing forms of authoritarian community. And how could such concepts do so when any such positing would inevitably reproduce just another version of that same authoritarian suppression? It is this last question which seems to be crucial, since if it is possible to think positing that is compatible (compossible) with the ontological dimension, without of course claiming to encompass it, then the door would be open to thinking an ontology of community that need not surrender valuation, and therefore affirm the political.

III

Taking off from the problem of the last section, what I will do in this section (and continue in the next), is to argue that within the same ontological perspective that Esposito is working from, different attitudes toward the person and community are both possible and necessary, and that some of these prove more amenable to political valuations and projects. In this section I specifically compare Esposito's discussion of the person with Agamben's, showing that the former's discussion of the theatrical *persona*, a key figure of his conceptual genealogy of the person, results in a "tragic" configuration tending to the impolitical, while Agamben's discussion of the same figure leads us to a "comedic" view in which the difference between person and concrete individual becomes the ground for a politics of expiation and of contesting the domination imposed by the apparatus of the person.[33]

Toward the beginning of the chapter devoted to the genealogy of the person in *Third Person*, Esposito presents the relation between the person and the living individual in terms of the changing configuration of the theatrical mask donned by actors on the Roman stage. What Esposito emphasizes throughout this discussion is that the theatrical mask becomes identified with the face of the actor who wears it. As he writes, "Originally understood as a stage garment or theatrical costume, *persona* also began to designate the individual who wore it."[34] This identification of the *persona* with the individual proceeds through adopting a fictional disguise, a mask (in this sense both the character played by the actor and the theatrical mask have exactly the same purpose), which comes to be "imprinted on the face of the wearer, until they correspond in every detail."[35] Esposito then shifts his attention from the stage to the funereal scene, in which, we are told, the correspondence between mask and face reaches the point of outright identification in the death mask worn by the deceased, replicating exactly the latter's facial features. The death mask serves to idealize the dead person, so that "in the face of death it was no longer the human being who donned a mask, but the mask that was incarnated in the human body, to the point of constituting its most authentic expression."[36] Far from maintaining the theatrical difference between the individual's face and the mask donned to designate the *persona*, the death mask closes that gap irrevocably, while at the same time idealizing the *persona* as the authentic or real face, the face of life. The reversal between the category of the person and the actual individual, wherein the latter is subordinated to the former, is then taken by Esposito as the paradigm case of the relation between person and individual in its long history in Western culture, as elaborated above.

The symbolic value of the death mask for Esposito's subsequent examination of the concept of the person cannot be stressed enough. The death mask, as an externally imposed idealization of the individual, is also the latter's occultation, conveying the individual's symbolic death in being engulfed by, and indeed sacrificed to, what is initially a prosthesis (a stage mask), which in a strange reversal takes on a life of its own only by expropriating that of its bearer in coinciding with it. That Esposito should choose the death mask as the point of reversal in which the gap between mask and wearer disappears is thus a potent symbol of the immunizing function of personhood itself, which, as already discussed, protects only by diminishing and ultimately sacrificing to death the life it is intended to protect. In the schema of immunization that Esposito develops, this is the point at which death takes over life by quite literally taking on life's features as its own: from this point onward, life will be lived under the shadow of the persona as death mask, that is, as permanently constrained by the latter's idealizing yet artificial features, an idealization that ultimately hides the fact of the real individual's death.

At this point we can compare Esposito's discussion of the *persona* with the one provided by Agamben in his essay on Dante's *Divine Comedy* in the book *The End of the Poem*.[37] The problem that Agamben seeks to address in his essay is why Dante affixed to this work the label of "comedy." In his analysis, Agamben initially contrasts the concept of guilt as it operates in classical tragedy and comedy. In the first genre, tragedy, a noble and just character is subjected to a terrible fate which would be deserved only by someone truly wicked, while in the second one, a blameworthy character comes to be exonerated or justified: "tragedy appears as the guilt of the just and comedy as the justification of the guilty."[38] Next, in a move that is clearly inspired by Walter Benjamin, Agamben introduces a crucial distinction between the concepts of natural and personal guilt.[39] In the course of its development, Christianity takes over the ancient world's idea of tragic guilt in its doctrine of original sin, whereby natural guilt is incurred by each individual (as a member of the species as a whole), regardless of the personal intent or actions involved. Here we can see recapitulated the ancient notion of the tragic punishment of the just made guilty by an inexorable fate. However, in marked contrast to the tragic conception, Christianity also allows that natural guilt is susceptible to justification through confession and good works on the part of the individual; these bear witness to a good will that inserts a difference between the redeemed individual and his or her natural guilt that in the eyes of God and the Church serves to render the former blameless.

These distinctions between tragedy and comedy and natural and personal guilt, according to Agamben, also profoundly impact the positions and subsequent cultural import of character and actor on the stage. Ancient theatre already seemed to have a sense of the stakes involved. For instance, the Stoic critics whose views Agamben also imputes to Dante, suggest in the tragic form a persistent conflation between "actor and character" such that there is a reduction of the difference between the actor on the stage and the role played. As Epictetus apparently put this, "remember that you are meeting a tragic hero – not an actor but Oedipus himself."[40] As this implies, in tragedy the difference between actor and character is recognized by the audience but makes no practical difference to the outcome of the events on the stage. In contrast to the tragic, the comic genre introduces a gap between the role played and the one playing the role, between the mask (the *prosopon*) and the face of its wearer: "The wise man is instead one who, accepting without discussion whatever 'mask' has been assigned to him by fate, *represents* his part and thereby refuses to identify with it."[41] This "wise man" associated with comedy maintains and affirms the distance separating himself from his role, and it is this difference which in comedy makes a difference to the events on the stage, allowing for the justification of the guilty.

At this point, like Esposito, Agamben transposes his discussion of the person from the stage to socio-political life. What Agamben emphasizes, however, is that for Dante the distance separating the actor (here the natural individual) from the person (the public mask and legal-moral personality) becomes the basis for a kind of salvific or redemptive view of life, one that might be described as "comic" rather than tragic. Dante in this regard follows the Christian viewpoint which acknowledges that "nature and person do not coincide in the creature," and yet simultaneously and in contrast to the ancient world's concept of the person, affirms "the antitragic distance between actor and 'person.'"[42] This makes it possible for Dante, under the sign of comedy, to maintain that it is human nature as it appears in the individual that is innocent, while it is the legal-moral person that is the true bearer of natural guilt. Recalling Agamben's distinction between tragedy and comedy in which the first affirms guilt and the second justifies or expiates it, it is possible to say that Dante's notion of comedy is nothing less than a project of removing or expiating guilt. The natural innocence of the individual's life is affirmed in direct contrast to the fatal guilt of the legal person. Where tragedy affirms the law's condemnation of the human individual as inevitable, because of the individual's identification with the mask or *persona*, comedy affirms the natural innocence of the individual by expressly endorsing the individual's difference from the legal persona.

While we might think of this difference in terms of a capacity to merely elude or evade the guilt imposed by the persona of the law, and thus make a comparison the content of which would be substantially similar to Esposito's notion of ontological "affirmation," I think it is possible to go further than this. The argument might be developed in the following way: in the distance between individual and *persona*, it is not the individual who bears or incurs guilt, for the latter is innocent, but it is rather the individual's immersion in fateful situations that create the appearance, mask, or *persona* of guilt. The law, in other words, rather than nature, assigns guilt. If so, then the notion of expiation takes on an explicitly redemptive and political hue. The emancipation of the individual from the guilt of law and its fatal history reverses the value assigned to the difference between these two elements, supplying in the process the difference between the tragic and the comic. This is quite evident in the way in which these two accounts of the theatrical *persona* provide a model for the legal category of the person. For Esposito the death mask is a trope characterizing the person and its strategy of immunity in a substantially identical way across time. As a political technology that splits the human being up into living individual and legal-spiritual entity, the person is thought of by Esposito as being able to subordinate and contain the difference between itself and the natural individual. As I have argued, the symbol of the death mask in this capacity is significant. The death mask stamps death in the image of life, all the better to conceal the fact that life is already dead. The difference between mask and individual, while recognized, is reduced to insignificance. In the terms of the analysis put forward by Agamben, we might say that Esposito espouses, through the dispositif of the person, a formally tragic view of the relation between person and individual, history and nature, in which difference is inevitably reduced to sameness. Although Esposito does not deny – he indeed recognizes – the difference between mask and actor, person and natural individual, for reasons discussed in the previous section, he does not want to affirm this difference as providing a political alternative to the project of immunity, which tirelessly carries on subordinating one to the other on the very basis of their difference.

The conclusion I would like to draw from this comparison of Esposito and Agamben on the concept of the person is that the immune paradigm has had to contend with another, an emancipatory or "comedic" view of the difference between person and individual. The comedic view recognizes and affirms the difference between person and living individual in order to rescue the latter from the former. On Agamben's account, at least from the late Middle Ages, there is already the recognition of what in modernity will become the reality of social division, the separation between the proper representation of living individuals as persons by power, and what those individuals might be in and for themselves. To put it another way, in modern society there is a decoupling of knowledge from power.[43] In this decoupling, the tragic vision of law in its pretension to know is blinded because the individual's representation as person by power is seen as no longer corresponding to the living individual the latter purports to capture. Such decoupling not only makes it impossible for power entirely to close the gap between *persona* and individual, but it is this impossibility which stages a political project of expiation, that is, of using the natural innocence of the human individual against the category of the person, of condemning in turn the fatality of the law that condemns.

IV

In the last section we hit upon an important source of political valuation, namely, the difference between, on the one side, the tragic fatality of law that seeks to assimilate the individual to the person on the basis of their very distance from it, and on the other side, the comedic expiation of the individual by affirming this very difference. What remains in this section is to link the political value of comedy to the political problem of ontology as discussed at the end of section II. It will be my claim here that the way to go about doing this is to show that the political value of comedy can indeed appear in the domain of politics as an ontological claim, and, further, that this is precisely what Nancy's more recent writing has attempted by way of espousing a link between democracy and an ontological community of being-with or sharing out (*partage*).

To show this, it is worth entertaining a brief comparison between Esposito and Nancy on the issue of community. This is worthwhile not least because Nancy's work is explicitly acknowledged by Esposito as the basis for his own reading of Heidegger. To get to the root of the contrast I want to make it is perhaps enough to quote the following passage from Nancy's relatively recent essay "The Truth of Democracy," written, significantly enough, in defense of the May 1968 uprisings in France on the occasion of their fortieth anniversary:

> The renunciation of every principal form of identification – whether it be borne by the image of a King, a Father, a God, a Nation, a Republic, a People, a Man, or a Humanity, or even a Democracy – does not contradict, quite the contrary, the exigency of identification in the sense of the possibility for each and every one to identify him or herself (or as people like to say today "to take up a subject position") as having a place, a role, and a value – an inestimable value – in being together.[44]

What is immediately evident here is that Nancy endorses the sort of vigilance against the closure of the community under some image or figure of finality in a manner very close to what we have found in Esposito. Yet we also find here an apparently "positive" claim that the ontology of being-with gives, shares out, or even mandates to each individual a singular value on the basis of being-together and within its distribution of shares. The question is whether such a distribution is politically neutral in the same sense as Esposito's account of this relation was. What Nancy intends by his notion of ontological sharing out is "the affirmation of a unique, incomparable, unsubstitutable 'value' or 'sense'" donated to each individual in their sharing out.[45] This unique and incomparable value would be very close to the meaning of the political value of comedy in that it would affirm the unsubstitutable nature of the individual as unique and singular against the legal *persona* and its compulsion to impose guilt, its attempt to distribute singularity according to the order of the proper. But since this singular being is only possible on the basis of the sharing that is being itself, singularity cannot be thought of as anyone's property, but rather as the property of distributed being. In his mention of place, then, Nancy is not intending the term as though it were a social placement imposed from without or given in advance, as it would be under the ordering of the proper. But if this is so, the being-together of community for Nancy cannot just indifferently expose us to the negativity of our ontological relation. It also does two more things: first, it gives or donates a unique sense to each of us, while second, on the basis of this unique sense as something only graspable as shared or relational, it enjoins us to become subjects, even political subjects, but of a kind different than would be expected on the basis of the metaphysics of the proper.

In his text entitled "Compearance" Nancy provides a discussion of what such political subjectivity would entail. He does this through a consideration of what would amount to a series of denials. Denial of membership in a given community, a denial that would be articulated and enforced in the political domain (and so also in the domain of law, having to do with the *persona* of the legal person), would inevitably have ontological ramifications, for this denial would also deny the shared existence of being-in-common. Such a denial, Nancy writes, "exists as a denial of existence [...] and the existence to which one denies a sharing is itself a denied existence"; but precisely to this extent, we can conceive political denial as producing a further effect, one that "reaches all existence, for it touches the *in* of the in-common."[46] The denial of existence, articulated politically in its multiple senses, touches the ontological domain of co-appearing, not just of distinct beings, but in and through them, of the sharing out of being itself. So the political act of denying belonging doesn't just stay on the "ontic" surface of political relations, but instead goes to the ontological domain and touches what is possible for existence as such and for each.

If the politics of the proper can touch the ontological, it must be possible for the ontological to also touch the domain of politics. Nancy suggests such a relation at the start of his paper by dwelling on the meaning of his French term "*la Comparution*" that also serves as its title. The term is derived from the archaic Scottish legal term compearance, in turn taken from the Old English verb "to compear." Such compearance, according to Nancy, bears the sense of an inexorable summons to court, a summons to appear in and before the court. But this court, as Nancy makes clear, is not one convened by the legal apparatus of the person or in the name of the proper. Rather, the exigency runs deeper, since, as Nancy states, "This is a simple judgment, without appeal; it is not subject to a superior law (*droit*) since it issues from that which precedes law."[47] What this exigency preceding law imposes on us is the obligation of the "in" of the in-common itself, in the form of a simple but powerful question:

> Have we done right (*droit*) by that which still has no right? Right by our existence itself – or since this word is subject to misuse in the singular, by our existences, by their community? Before this law without law we have never ceased to compear. *In the end* we compear there naked.[48]

The law without law imposes on us a responsibility before which we stand naked and defenseless: to do justice or to do right by the "in" of the in-common. This injunction constitutes an ontological pressure on the political that allows us to see why Nancy believes that in the absence of the metaphysics of the proper there remains the necessity of subjectivity with respect to this compearance. So in some sense, for Nancy, the community of the in-common – *pace* Esposito – demands (political) subjects, and even calls them into being.

To what end? We might answer this question with a catch-phrase: compearance implies compossibility, since it is composed of nothing but the co-appearing of co-constituted singularities. The community of co-appearing gives rise to a politics of plurality and pluralization, of the maximal inclusion of singularities. Such compossibility stands against (in the political sense of taking a position), the proper that constitutes political identities by turning the unique and irreplaceable value of individuals (as singularities) into fodder for the affirmation and imposition of a fixed essence. Since all configurations of the proper will do this, such standing against will also have to be directed not only against the proper but even against that which is against the proper, at the point where the latter risks hardening into its own form of essential identification. No doubt this politics will be vertiginous, even seeming to court madness, since in the name of inclusion it will require a certain amount of identity-fixing of its own, and this will inevitably produce some exclusion and inequalities of shares. But this paradoxical situation should not distract from the guiding principle at stake, namely, that those modes of political subjectivity that seek to do justice to the compearance and co-constitution of the world are to be defended and advanced in the political arena against those which seek to categorize, fix, and exclude on the basis of a notion of the metaphysically proper.

If Nancy is not shy about affixing such a political orientation or strategy with the label of "democracy," he is equally clear that "democratic politics refrains from laying claim to this sharing (out), even though it guarantees its exercise."[49] Such guaranteeing of democratic exercise would suggest activity in the public sphere on the part of political subjects seeking to defend and promote the compossibility of the democratic sharing out of being, even as this political exercise demands a difficult but necessary restraint. This restraint that refrains from laying claim to sharing out, to imposing on democratic community a finalized and determinate form, is precisely the attempt to do justice to the law of compearance; in other words, compearance demands stopping short of imposing the contours of the community in such a way that does not become the establishment of a new proper, a new finite order of political relations that once again acts as a hard boundary for exclusion and the denial of ontological sharing-out of common singular-plurality. Democracy, for Nancy, is the best political order not simply because it is preferable to the others but also because it alone is capable of explicitly acknowledging that the human is not given once and for all, that it is always produced by the unlimited sharing-out of sense according to the law of equality of our co-belonging to and for each other.

From this it should be evident how the discussion of the difference between the individual and the person outlined in the previous section dovetails with Nancy's discussion of compearance and democracy. There is in Nancy's work, in other words, an ontological exigency that prescribes a normative political value to the project of emancipation of singularity from the hierarchies dominated by the legal *persona* and the proper. So the historical movement that transmutes a formal regime of tragic condemnation into one of comedic liberation is matched in Nancy's account of democracy by a certain normative dimension of ontological exigency that is emancipatory and which situates an emancipatory politics, one that simultaneously eschews emancipation as a realized "state" or finalized end.

On reflection, one has the clear sense that Esposito could have said something like this but has chosen not to. But the question remains why Esposito is unwilling to affirm the emancipatory meaning of the in-common we find Nancy unfolding in the direction of a radical or infinite democracy.[50] As I have tried to suggest, the similarities in their approach follow from their shared appropriation of Heidegger. Yet this shared appropriation of Heidegger also causes them to diverge precisely to the extent that Esposito seeks to use Heidegger's radical deconstruction of community against Heidegger's own endorsement of an essential community situating a political project of realization. Nancy is less concerned with this project of reading Heidegger against himself if only because he seeks to develop *mitsein* into a "first philosophy" that would simply bypass Heidegger's philosophical inconsistencies, becoming in the end a quite different project.[51] In contrast, in Esposito's sparring with Heidegger, there is a wariness that seems to impose an internal limitation. This wariness is conceived around the drive to closure and the nothingness of relation that keeps it open. In tandem with such wariness there is the question of narrative and history. Because Esposito opts to trace the history of community from a determinate communal form – that of ancient Rome – he apparently embraces a formal narrative of community as loss of origin and movement toward monolithic closure, even if his stringent deconstruction of community staves off the usual content of such a narrative as the nostalgia for return. But as I've also tried to show, this seems to turn Esposito's history toward the tragic, leaving unformulated the emancipatory potentials that unfold out of that history itself, as tragedy gives rise to comedy and the republican project is displaced by the democratic.

notes

I would like to take this opportunity to express my thanks to Greg Bird for his very helpful comments on an earlier draft of this paper, as well as to my former colleagues at the (now defunct) Theory Department of the Jan Van Eyck Academie in Maastricht, the Netherlands.

1 Jean-Luc Nancy, "Responding for Existence," *Studies in Practical Philosophy* 1.1 (1999): 2–11 (2).

2 This is how Esposito articulates the difference between his work and Nancy's in the paper in this special issue entitled "Community, Immunity, Biopolitics."

3 To this extent I seek to blur the rather stark alternative that Peter Gratton seems to see in his essay in this issue between immanence and transcendence in the work of Esposito (who models an "affirmative biopolitics" on the great "philosophers of immanence," Spinoza and Deleuze) and Nancy. From this perspective it must be kept in mind that Nancy himself refuses the choice between immanence and transcendence as ultimately impossible. See Jean-Luc Nancy, "Imm/Trans," *Polygraph* 15/16 (2004): 11–12.

4 Roberto Esposito, *Communitas: On the Origin and Destination of Community*, trans. Timothy Campbell (Stanford: Stanford UP, 2010) 6.

5 Ibid.

6 Roberto Esposito, *Third Person*, trans. Zakiya Hanafi (London: Polity, 2012) 83.

7 Ibid. 82.

8 Ibid. 77.

9 Ibid. 79.

10 This discussion of the difference between the person and human being recalls Agamben's similar concept of the state of exception as a mobile and shifting threshold between inclusion and exclusion. As Timothy Campbell points out in his article on Esposito's notion of the impersonal, the difference between Agamben and Esposito on this point is that for the latter the zone of indistinction passes through and is located in the entity who is both human and (potential) person, rather than situated between the separation between *zoe* and *bíos*. See Timothy Campbell, "'Enough of a Self': Esposito's Impersonal Biopolitics," *Law, Culture and the Humanities* 8.1 (2012): 31–46.

11 Hannah Arendt, *Origins of Totalitarianism* (New York: Harcourt, 1973).

12 *Third Person* 92.

13 Ibid. 101.

14 Ibid.; one thinks here of J.S. Mill's contention at the beginning of *On Liberty* that despotic government is entirely suited to those he calls barbarous peoples.

15 *Communitas* 15.

16 Ibid. 3.

17 Ibid. 2.

18 Ibid. 16; original emphasis.

19 Ibid. 99.

20 Ibid. 92.

21 Ibid. 94.

22 Ibid. 93.

23 Ibid.

24 Ibid. 95.

25 Ibid.

26 Ibid. 96.

27 Ibid. 97.

28 Ibid. 102; original emphasis.

29 Ibid. 14.

30 Ibid.

31 Esposito's concept of the impolitical is therefore similar to Schmitt's concept of the *katechon*, the "delayer" who does not redeem but merely holds open the place of transcendence by delaying the triumph of the anti-Christ. See Carl Schmitt, *The Nomos of the Earth in the International Law of the Jus Publicum Europaeum*, trans. G.L. Ulmen (New York: Telos, 2003) 59–60. For Esposito this function of delay (perhaps via the alteration of difference within identity) is indefinite: there is no "messiah" who is to come, no eschatology, only the infinite task of preventing the worst.

32 This problem of ontology's connection to the political is the basis for both Badiou's philosophical project of recasting politics as one of the fundamental domains of ontology (along with those of love, science, and art), and for the return to Spinoza's creative ontology as found in Negri, Deleuze, and Althusser.

33 This presentation of Agamben as potentially more affirmative of politics is somewhat ironic given that he does not always seem to be so. For example, in apparent contrast to the position he takes in this discussion of Dante, in the appendix to his book on *The Coming Community* (trans. Michael Hardt (Minneapolis: U of Minnesota P, 1993) 92) Agamben writes that "The world of the happy and that of the unhappy, the World of the good and that of the evil contain the same states of things; with respect to their being-thus they are perfectly identical." This would seem to license the same ontological indifference toward how things are in the world, and so would appear to endorse a similar remove from the political world of valuation and contestation – whether this is actually the case would require a broader examination of Agamben's work that is beyond the scope of my discussion here.

34 *Third Person* 74.

35 Ibid.

36 Ibid. 75.

37 Giorgio Agamben, "Comedy" in *The End of the Poem: Studies in Poetics*, trans. D. Heller-Roazen (Stanford: Stanford UP, 1999) 1–22.

38 Ibid. 8; original de-emphasized.

39 There are several texts by Benjamin that one might cite here, but perhaps the most obvious one is the "Critique of Violence," trans. Edmond Jephcott, in *Selected Writings*, vol. 1 (Cambridge, MA: Harvard UP, 1999).

40 Quoted in Agamben, "Comedy" 17.

41 Ibid. 17–18; original emphasis.

42 Ibid. 19.

43 This distinction is Claude Lefort's; see Lefort, "The Question of Democracy" in *Democracy and Political Theory*, trans. David Macey (Cambridge: Polity, 1988) 9–20.

44 Jean-Luc Nancy, "The Truth of Democracy" in *The Truth of Democracy*, trans. Pascale-Anne Brault and Michael Nass (New York: Fordham UP, 2010) 1–34 (27).

45 Ibid. 24.

46 Jean-Luc Nancy, "Compearance," trans. Tracy Strong, *Political Theory* 20.3 (1992): 371–98 (392).

47 Ibid. 372.

48 Ibid.; original emphasis.

49 "The Truth of Democracy" 17.

50 There are certainly passages in Esposito's work where he does situate politics and community in a more affirmative and emancipatory light and which remain consistent with his philosophical project. For instance, in his essay "Freedom and Immunity" Esposito writes that "freedom is the internal exteriority of community: the part of community that resists immunization, that is not identical to itself, and that remains open to difference" in *Terms of*

the Political: Community, Immunity, Biopolitics, trans. Rhiannon Noel Welch (Fordham: Fordham UP, 2013) 56.

51 See François Raffoul, "The Logic of the With: On Nancy's *Être singulier pluriel*," *Studies in Practical Philosophy* 1.1 (1999): 36–52, for an explication of this.

lorna weir

ROBERTO ESPOSITO'S POLITICAL PHILOSOPHY OF THE GIFT

Deconstructive philosophy has challenged close to a century of philosophical and anthropological work theorizing the gift as a form of exchange. Derrida's revolt against conceptualizing the gift in terms of exchange/ economy was done in the name of saving the gift from the metaphysics of the proper, severing the association of the gift with property, appropriation and all calculated exchange. Roberto Esposito has extended the deconstructive theory of the gift into political philosophy, theorizing the gift as the transcendental form of political obligation. He argues that the political obligation to give constitutes the being-in-common of community. Esposito's philosophy of what might be called the "political gift" marks a distinctive contribution to what has now become a global discussion of community and the common (Nancy).

This article begins with an overview of Esposito's theory of gift from his initial engagement in the journal *La Revue du MAUSS* (1995) to his theory of the *munus* beginning with the publication of *Communitas* (1998) and continuing to date. I thematize Esposito's relation to Maussian gift theory, most particularly Marcel Mauss's *The Gift*, emphasizing his distance from Derrida's treatment of Mauss. In Esposito's philosophy of *communitas*, the *munus* consists of the single obligation to give, a logic of donors without receivers. Yet the *munus* additionally establishes relations of reciprocity, mutuality, and gratitude. I argue that reciprocity, mutuality and gratitude are not logically possible in a gift system where the donor has no other and thus the gift no recipient. The inconsistencies in Esposito's treatment of the gift may be addressed by distinguishing two forms of gift in *communitas*: the impersonal gift characterized solely by the obligation to give to the gods/*communitas* (Mauss's little known fourth obligation) and the reciprocal gift characterized by the obligations to give, receive and return (the gift) (Mauss's first three obligations). I additionally suggest that Esposito's philosophy of *communitas* would be strengthened by attending to the symbolic element in the *munus* as political gift. The *munus* as impersonal gift operates as a vertical relation of expenditure made possible by the collective representation of the political we, which is that to which and for which the *munus* is given. As impersonal gift the *munus* is a symbolic practice through which the members of *communitas* represent to themselves their being-in-common. In contrast, the reciprocal gift constitutes social ties and networks that establish the space of a differentiated communal life distinct from the being-in-common of the *munus* as a political relation. In this way the distinction between the impersonal and the reciprocal gift gives rise to a minimal division between the political and the communal within the transcendental structure of *communitas*.

esposito's political gift

Esposito's first sustained comments on the gift, "Donner la technique," surprisingly appeared in *La Revue du MAUSS*. The choice of journal surprises given Esposito's self-identification with deconstructive philosophy and Derrida's relentless hostility to Maussian gift theory in his 1991 book *Given Time*, which had appeared four years prior to the publication of Esposito's article. The *Revue* takes the name of MAUSS, an acronym for "Mouvement anti-utilitariste dans les sciences sociales." The leading international journal and network of scholars organized around the work of Marcel Mauss, the *Revue* has opposed the presence of "economism in the social sciences and instrumental reason in moral and political philosophy."[1] The publication of Esposito's article on the gift in *La Revue du MAUSS* (under the heading "Autour du don") is a gesture of goodwill and dialogue across opposing schools, one social scientific, the other philosophical, at the very site of their disagreement. The occasion called for a diplomatic gift.

"Donner la technique" reads the history of gift theory from Mauss to Derrida through three "keywords" that Esposito argues have characterized its history: critique, constitutive ambivalence, and origin. In his first writing on the gift, Esposito diverges from Derrida in situating Maussian gift theory as a critical project in alliance with deconstructive gift theory rather than as an opponent from whom the gift must be wrested.

Esposito's outline of gift theory as critique begins with a short summary of the intervention that those grouped around the contemporary social science school MAUSS have undertaken, exemplified by Alain Caillé's *Critique de la raison utilitaire* (1988). He treats Maussian thought as a critique of Western modernity in its refusal of utilitarianism and instrumental reason. These latter combine to form a "productivo-calculating attitude" crosscutting economics and politics, an attitude that seeks to maximize the productivity of the means–ends relation (190). Esposito (190–91) calls this attitude "la technique": a globalizing and destructive form of reasoning that is predicated on the idea that each act has a cause and underlying interest. The Maussian school breaks with this calculated governance of nature and humanity, Esposito (191) maintains, through its attentiveness to the gift as the basis of the social tie and of our sociability, a place of being together that is not founded on the calculations of technical governance (*la technique*). Esposito distances himself from Derrida's interpretation of Mauss as the creator of a counterfeit gift theory which encompassed the gift within the metaphysics of the proper. In contrast, Esposito frames both MAUSS and the Maussian theory of the gift as a critical stance opposed to *la technique*.

"Constitutive ambivalence" forms the second characteristic of Maussian gift theory named in "Donner la technique" (195). The first paragraphs of Mauss's *The Gift* had outlined a program that famously conceptualized its object of study as double. Examining "exchanges and contracts that take the form of presents," Mauss observed that "in theory these are voluntary, in reality they are given and reciprocated obligatorily" (*The Gift* 3). The double structure of the gift has remained an abiding theme in the theory of the gift since Mauss. The gift has doubles of many kinds: apparent voluntariness that is also obligatory (Mauss); expansiveness such that nothing is not gift and thus gift is nothing (Caillé); generosity so destructive/expansive that holding onto wealth appears miserly (Bataille). Esposito (197–99) provocatively situates Derrida's "impossible gift" within the Maussian gift tradition as a development of Bataille's work, rather than, as one would surmise from Derrida's sustained critique of Mauss, a break from that tradition. The discussion of the gift's "constitutive ambivalence" finishes with Heidegger's philosophy of the event (*Ereignis*), which is a giving without a gift, the appearance of the possibility of giving a gift (199). Esposito incorporates Derrida and Heidegger as members of the patriline theorizing the ambivalence of the gift.

"Origin" is the last of the three themes that Esposito identifies as characteristic of Maussian gift analysis: the gift as disguised trickery, the poisonous gift (the most famous discussion of which is Mauss's "Gift-gift" (199). The gift, Esposito writes, has its origin in the myth of Prometheus, who gave the gift of technique to humanity, intending to free humans of their fear of death. A tragic gift, humans use technique to challenge the gods, who are then pressed into resolving the problem with one final gift to humanity: politics. The gift of politics enables humans to shape technique more wisely and to remember what technique forgets.

Esposito's later work has pursued the theme of the relation between the gift and the formation of the political introduced in his 1995 article. From "Donner la technique" to the trilogy of *Communitas*, *Immunitas*, and *Bíos*, Esposito treats the gift as establishing a political horizon beyond the logic of instrumental rationality and the maximization of productivity. However, with the publication of *Communitas* in 1998 he broke with his earlier formulation of politics through Ancient Greek mythology as a gift of the gods bestowed to limit the destructive power of technique. The formation of the political, no longer externally given, is interrogated and conceptualized in *Communitas* as our being-in-common constituted via a particular form of gift: the *munus*. Esposito's tangential remark about the relation between the gift and the common made in the context of his 1995 discussion of critique would come to the fore in *Communitas*. The placing of deconstructive gift theory in historical continuity with Mauss, however, was an exercise that Esposito has not repeated.

gift and communitas

The publication of *Communitas* in 1998 marked a major transformation in Esposito's approach to the gift, which he reconceptualized in relation to his emerging philosophy of community. The gift constitutes and renews community by conjoining its members in the obligation to give, but not to receive. I argue that Esposito's conceptualization of *communitas* is inconsistent on the question of whether or not it is characterized by reciprocity (cycles of giving and receiving that constitute ties) or bound to a non-reciprocal logic of giving without receiving. I suggest that Esposito's theory of the gift is characterized by an unresolved tension between the impersonal gift that founds the political in its verticality and the horizontal, reciprocal gift relations between social actors (individual and collective).

Communitas is intrinsically characterized by obligation and indebtedness: "*communitas* is the totality of persons united not by a 'property' but precisely by an obligation or a debt" (*Communitas* 6). The members of *communitas* are linked through a particular type of gift, the *munus*, which binds them together "by the same duty or gift to give" (*Terms of the Political* 39). Turning to the Latin etymology of community in the "Introduction" to *Communitas*, Esposito directs attention away from its prefix, *com*, and towards *munus*, which in Ancient Rome was associated with *onus* (task), *officium* (office), and *donum* (gift) (*Communitas* 4). The *munus* is an obligation to give goods and services and more generally the self, but with no reciprocal obligation to receive; it is an obligation that can in principle never be fulfilled: "Although produced by a benefit that was previously received, the *munus* indicates only the gift that one gives, not what one receives. All of the *munus* is projected onto the transitive act of giving" (5). This analysis simultaneously acknowledges but elides the constitutive position of the *munus* as a response to a prior gift, an obligatory *return* of the gift; without the antecedent gift, the *munus* would not be.

In requiring of its members the obligation to give, community expropriates their identities, revises the boundaries between the private and the common, and thereby generates conflict. *Communitas* is formed in the "expropriation" of identity and its transformation in ceaseless obligatory giving. In his more recent book, *Terms of the Political* (2013), Esposito, citing Heidegger, specifies the duty/work/gift that constitutes community as care, the care that frees the capacity for care in the subject (25–26).

Esposito's discussion of *munus* cryptically references its role in Ancient Roman history. He is concerned principally with the *munus* in Roman public life and its attendant duties rather the *munus* in private life. In the Ancient Roman world public *munera* (plural of *munus*) were obligations/gifts owed to the municipality/city and the central administration. Residents and citizens of a city were considered to benefit from their membership in it and to have resulting moral and legal obligations (*munera*) to contribute services and wealth in return. Personal *munera* (*munera personalia*) consisted of non-financial services, such as physical labor (especially military service and service as a magistrate) and overseeing municipal offices for grain and water. Financial *munera* (*munera patrimonalia*) consisted of financial and other obligations, including taxes, placed on municipal residents in relation to the property they held in the city. Owners of real property (land and buildings) were also obligated to care for city streets and provide both horses and housing for soldiers (Corbier). Tangentially I note that in Ancient Rome the word "*munera*," without further qualification, had the primary meaning of public gladiatorial combat.

In the *Communitas* trilogy Esposito takes the Ancient Roman public *munus* as the exemplar of communal obligations that must be fulfilled: the gift that each must give. The *munus* has the feature of being mandatory: "this is a gift that one gives because one must give and because one cannot not give" (*Communitas* 4). Esposito is emphatic that the *munus* has the quality of being a unidirectional, asymmetrical gift donated without the expectation of return; being-in-common is established through giving, not receiving: "[t]he subjects of community are united by an 'obligation' in the sense that we say, 'I owe *you* something,' but not, 'you owe *me* something'" (6). The *munus* places the donor in the position of Mauss's first obligation (the obligation to give). Yet, as noted above, *munus* is "produced by a benefit that was previously received" (5). The members of *communitas must return* the gift; the *munus* (implicitly) recognizes an anterior indebtedness. The *munus* thus has a structure of receiving-and-giving rather than unidirectional giving. Reciprocity would appear to be intrinsic to the *munus* as receiving-and-giving, with the proviso that the circulation does not necessarily involve interdependence among the members of community.

To be sure, Esposito's work on *communitas* has taken a foregrounded critical stance against theorizing community in terms of interdependence and reciprocity; his contribution to the present collection continues this critique. These explicit statements exist in tension with passages that link *communitas* with reciprocal relations:

> [*I*]*mmunitas* is not just a dispensation from an office or an exemption from a tribute, it is something that interrupts the social circuit of reciprocal gift-giving, which is what the earliest and most binding meaning of the term *communitas* referred to. If the members of the community are bound by the obligation to give back the *munus* that defines them as such, whoever is immune, by releasing him- or herself from the obligation, places himself or herself outside the community. (*Immunitas* 6)

> If *communitas is* what binds its members in a commitment of giving from one to the other, *immunitas*, by contrast, is what unburdens from this burden, what exonerates from this responsibility. (Esposito, "Community, Immunity, Biopolitics" 84)

In these passages communitas is identified as a space of reciprocal gift relations in contrast to suspension of reciprocity in *immunitas*. The broader significance of this seeming instability in the concept of *communitas* requires exploration.

Although Esposito centrally engaged with the work of Mauss and the Maussians on the gift in his 1995 article "Donner la technique," his later work contains only glancing references to

Mauss. Maussian gift theory is nonetheless more conceptually significant in his writings on the philosophy of community than it had been in his previous work on technique and politics. From the publication of *Communitas* in 1998 Esposito's political philosophy has been consistently silent on Mauss's formulation of the social obligation and the gift cycle, while at the same time using its analytic terms such as obligation, reciprocity, debt and gratitude. Mauss argued that gift exchange operates in three cycles: the obligation to give, to receive, and to return the gift.[2] These obligations operate between collectivities, representatives of collectivities, persons or any combination of these.[3] The obligation to give (the first obligation) is simply to give the gift; the gift retains an element of the donor, reminds the recipient of the donor. Once the gift has been accepted (the second obligation), the recipient is required to perform a ritual expression of gratitude, becomes indebted to the donor, and must return the gift (the third obligation).[4] The obligatory structure of the gift constructs reciprocal relations, with gifts circulating to form and renew social ties both vertically and horizontally; the gift cycle constitutes social reciprocities.

The law of *communitas* consists solely of the first Maussian obligation: the obligation to give. However, in Esposito's account of the *munus* the obligation to give is conceptualized as a position of indebtedness (characteristic of the second obligation), which would imply the donor has received a gift, although it is not clear how this is possible in a gift system which obligates solely giving. There can be no indebtedness without alterity; indebtedness minimally requires a debtor and a subject to whom she is indebted. Without an alternative and non-Maussian conceptualization of indebtedness, one reverts to the Maussian position that indebtedness characterizes the recipient of the gift, not the donor. A gift system consisting solely of donors thus has no indebtedness unless the primary social debt is thought to be owed to those whose care and protection have provided the conditions for our existence. The latter debt is alluded to in Esposito's discussion of the Christian *munus*, but he does not provide an argument there for how the social and political indebtedness of the donor arises, solely the debt to the Christian god. If humans have a generalized debt relation to each other, the question of to whom/what they owe the debt could be conceived variously as kin, the earth, the god-term, or the political. Below I argue that the *munus* takes the form of Mauss's fourth obligation, the obligation to the gods and their representatives, understood in a Durkheimian sense as the collective representation of the being-in-common of *communitas*.

Gratitude is also a feature of the *munus*, although the locus/subject from which gratitude arises is indeterminate: "[t]he *munus* is the obligation that is contracted with respect to the other and that invites a suitable release from the obligation. The gratitude that *demands* new donations" (*Communitas* 5). Again, gratitude is a ritual performance binding on those who are to return the gift, not the donor. It characterizes the second obligation (to receive the gift) rather than the first obligation (to give the gift). Conceding that he is "forcing slightly the concept of 'gratefulness,'" Esposito reasons that the donor:

> [u]nequivocally "owes" something of which he was the beneficiary and that he is called to acknowledge in a way that places him "at the disposition of" or more drastically "at the mercy of" someone else? What dominates in the *munus* is, in other words, reciprocity or "mutuality" (*munus-mutuus*) of giving that assigns the one to the other in an obligation (*impegno*). (Ibid.)

In addition to indebtedness and gratitude, the *munus* is marked by processes of reciprocity and mutuality. In the passage cited immediately above, the donor is a beneficiary who returns the gift; the giver "cannot not give" because s/he has received of the gift from another to whom s/he is the beneficiary (and thus indebted). The *munus* thus has a more complex system than that of a donor bound to the imperative of giving; instead, the donor additionally occupies the position of recipient to whom a more powerful donor has given a gift and to whom the recipient both displays gratitude and wishes to return the gift. Nonetheless, the logical possibility of reciprocity and mutuality would be precluded in a gift system that contains only the obligation to give. There initially appear to be three obligations rather than one at play in the *munus*: the obligation to give, the obligation to receive, and the obligation to return the gift.

The *munus* takes a dyadic form where the donor gives to "the other" and the resulting relation between donor and recipient is characterized by reciprocity/mutuality. But who is "the other" in this transcendental gift system consisting only of donors? To whose mercy is the donor/recipient exposed? How does mutuality predominate under conditions where the donor may be "at the mercy" of the other? These are significant concerns in a political philosophy where *munus* is a component in a wider theory of political obligation.

Esposito's philosophy of the *munus* as a transcendental gift system would thus appear to have two inconsistent levels, each with a Maussian lineage. The first and explicit level theorizes *munus* as the obligation to give which is binding on all members of community, creating being-in-common. At this level, the *munus* as obligatory gift is conceptualized uniquely through Mauss's first obligation (the obligation to give). At the secondary and implicit level, the *munus* grows into a system of indebtedness, gratitude and reciprocities that requires all three obligations; there can be no reciprocity without others who return the gift. Community confined to the first obligation lacks indebtedness and reciprocity.

The transcendental gift system of *communitas* is characterized by a tension between what might be termed the impersonal gift and the reciprocal gift. Consistent with Esposito's analysis, the impersonal gift consists of a unidirectional obligation to renew the being-in-common that is a condition of our existence. The obligation to give but not to receive that is found in the impersonal gift does not result in the formation of reciprocal social ties among social actors (the reciprocal gift). Rather, as I will discuss more fully in the final section of this article, the impersonal gift constitutes transcendental *communitas* in its vertical obligation to the political. The impersonal gift has the structure of Mauss's little discussed fourth obligation, the obligation to give to the gods and their representatives, where, following Durkheim, the gods (like *communitas*) may be taken as symbolically constituting the collective unity of group which represents their being-in-common. The anthropologist Maurice Godelier (30), criticizing Mauss, suggests

> [h]e should also have taken into account that the gods are free to give or not, and that men approaching the gods are already in their debt, since it is from them that they have already received the conditions for their existence. His analysis fails to take into consideration the fact that the gods and the spirits are already superior to humans, and that the human donors are from the outset inferior to the godly receivers.

Those bound by the non-reciprocal duty of the *munus* to give to *communitas* are in its debt and inferior to *communitas*; they are bound to give gifts to *communitas* in the manner of those in the Ancient Roman world who gave *munera* to the cities of which they were members, cities which were superior to any resident or corporate group within them. The impersonal gift, imperative and unilateral, renews the being-in-common that is the condition for and superior to all reciprocal gift relations constituted through the cyclical obligations to give, receive and return the gift.

the *munus* and deconstructive philosophy

In an endnote to *Communitas* (fn. 23, 153), Esposito distinguishes between a "deconstructive" line of scholarship on the gift and a "constructive" one organized around the journal *MAUSS*. His own work on the *munus* and Derrida's on the impossible gift are deconstructive contributions to the theory of the gift, but while Esposito fashions a political philosophy of community around the being-in-common generated by the gift, Derrida undertakes an ontology of the gift. They differ also in their relations to Maussian gift theory, although their common divergence from Mauss is more complex than might initially appear.

Esposito and Derrida employ the gift to refuse the metaphysics of the proper and its associated metaphysics of the subject. As Greg Bird writes (this issue), the proper derives from *proprius* (adjective) as one's own, individual, or characteristic, particular. The derivates of *proprius* include property, appropriation, expropriation and a wide range of other Euro-american property distinctions. The subject form that coincided with the proper has had the qualities of autonomy, fixity and self-identity over the course of a lifetime: the proprietary subject. But the *munus* breaks with the metaphysics of the subject because it is not a property relation, "[i]t [the *munus*] isn't having, but on the contrary, is a debt, a pledge, a gift that is to be given, and that therefore will establish a lack" (*Communitas* 6). In *communitas* the subject constantly gives herself away, "driven by an unavoidable obligation because it is one with the subject's own proper desire" (18). *Communitas* as a giving up of fixed self-identity strikes down the conditions that support the proprietary subject.

The *munus* follows the path of the impossible gift in its resistance to the metaphysics of the proper. The gift, Derrida writes in *Given Time*, interrupts economy as property, allocation and the law of apportionment, together with the circulation of goods, services and money (*Given Time* 6). Derrida dissociates the gift from economy (*oikonomia*) as such, treating the gift as incalculable and therefore beyond property, appropriation and profit. When the circular representation of time found in *oikonomia* and the history of metaphysics becomes culturally dominant, "the gift is impossible" (9). The gift-beyond-economy breaks off at a tangent from the circle of *oikonomia*, interrupting its circulations and instantaneous present temporality in a movement that "tears time apart" (ibid.).

The philosophy of the impossible gift speaks in allusive riddles. No subject is requisite to this gift, which is neither given nor received by anyone, nor does what is given take the form of a thing or act; the impossible gift suspends the metaphysics of the subject by eliminating the subject as such. The consciousness, memory and anticipation of the gift are precluded by "radical forgetting." Turning to Heidegger's discussion of *es gibt* (literally "it gives," but translated as "there is/are"), Derrida links the impossible gift to metaphysics, treating Being and Time as given, as gifts: *es gibt Sein* and *es gibt Zeit*. The impossible gift has the structure of Being and Time (Diprose 68; Jenkins 85). The gift is an *Ereignis*, an event that leads beyond the repetition of the same, an event in a narrative; event and narrative are historical, with the impossible gift as the unforeseeable exception governed by the conjunction of chance with intentional freedom (*Given Time* 123–25). Derrida refers to his analysis of the impossible gift as "a little mad" (35) and repeatedly remarks that the impossible gift may not exist. The gift is being removed from Maussian hands for the world-historical purpose of exiting the proper; the *angelus novus* arrives.

Esposito's approach to deconstructive gift theory is not through an ontology of the gift but through a transcendental political philosophy. He conceptualizes community as "the transcendental condition of our existence, given that we have always existed in common" (*Terms of the Political* 14). His philosophy of community does not posit the *munus* as a determinate gift system but as the condition of possibility for all determinate gift systems. The *munus* founds community through the obligation to give; unlike the impossible gift, the *munus* is given by a donor, a subject that gives away rather than acquires, a subject in relatedness rather than abstraction from the other. Where Derrida resolves the metaphysics of the subject through the elision of the subject from the gift, Esposito reconstructs that subject as one who desires the obligation of giving to the other. The donor of the gift is a subject whose identity is revised through her giving to others in a continuing process of expropriation. If everyone must give, community has no borders and no fixity, or, as Esposito puts it, the community has nothing in common.[5] Esposito and Derrida thus address the metaphysics of the proper and subject in differing ways, Esposito through theorizing a transcendental community of the gift lacking the proper and the proprietary subject, and Derrida through an ontology anterior to economy and the subject.

Derrida and Esposito depart from Maussian gift theory in differing directions. From the outset the *munus* is clearly no impossible gift as it is binds the political subject to Mauss's first obligation (the obligation to give), which forms the law of community. In this way the transcendental community retains a subject: the donor. Political *communitas* begins in the single, austere obligation to give but, as I have shown above, soon the little Maussian categories of gratitude, debt, reciprocity come to take up residence in community and both horizontal and vertical solidarities multiply. In contrast, *Given Time* begins with a bonfire of Maussian analytics. Derrida, arguing in what he calls a "peremptory" manner (*Given Time* 13), asserts that "[f]or there to be a gift, there must be no reciprocity, relation, exchange, countergift or debt" (12). The first chapter of *Given Time*, "The Time of the King," in effect argues that Maussian gift theory annuls the gift by assimilating it to the metaphysics of the proper. Given that framing, all elements of Maussian analysis must be removed from Derrida's deconstructive philosophy of the gift; the impossible gift is devoid of obligation.

Derrida's relation to Mauss would initially seem to be one of critique, displacement and substitution. After the implicit critique of *The Gift* sketched above, Derrida then turns to an explicit critique in a review to end all reviews. Mauss is accused of failing to theorize his object, which then leads him to misrecognize the word "gift" (and its translations) for the gift.[6] "How," Derrida asks (*Given Time* 25), "does he [Mauss] produce 'the gift' as a category that occurs across cultures?" Derrida, however, does not substitute the impossible gift for the Maussian gift in a movement of philosophical triumph over anthropology. He states that his intent is *not* to argue "there is no exchanged gift. One cannot deny the phenomenon, nor that which presents this precisely phenomenal aspect of exchanged gifts" (37). To deny the phenomenon of the exchanged gift and substitute the impossible gift would render the logic of the determinate gift unintelligible. Instead, Derrida argues that "two major structures of the gift" exist:

> There would be, *on the one hand*, the gift that gives something determinate (a given, a present in whatever form it might be, personal or impersonal thing, "natural" or symbolic thing, thing or sign, nondiscursive or discursive sign, and so forth) and, *on the other hand*, the gift that gives not a given but the *condition* of the present in general, that gives therefore the element of the given in general. (54)

The gifts of Being and Time are the conditions of determinate gifts in general. Derrida identifies the impossible gift as a moral good, the good of giving in itself, giving prior to any gift (38). Nonetheless, the impossible gift does not annul the determinate gift; instead, the gift has two forms, the impossible and the determinate. Derrida does, however, contest Mauss's terms of analysis, wishing in particular to loosen the relation between the determinate gift and exchange through the theory of narrative.[7]

Esposito and Derrida share a debt to Mauss, but where Derrida warrants Maussian analytics (in revised form) for theorizing the determinate gift, Esposito applies Maussian categories to the transcendental gift system of political community. In this sense the gift theories of Esposito and Derrida have a fundamentally different relation to Maussian gift analysis. In Esposito's political philosophy, the transcendental gift system is conceptualized through a series of Maussian gift categories: the obligation to give, reciprocity, indebtedness and gratitude. In Derrida's thought the Maussian analytics of the determinate gift and the ontology of the impossible gift exclude each other, with the line of separation/conjunction to be investigated.

the impersonal gift, the symbolic and the political we

A gift system that has only the obligation to give contains donors, but no recipients, no gratitude, no debt and no reciprocity; the *munus* verges on the impossible. How is reciprocity logically possible where actors are bound only to the obligation to give and not receive? To whom/what is the *munus* given such that reciprocity arises? I suggested above that the inconsistencies in Esposito's theory of the *munus* be addressed by distinguishing two forms of gift in *communitas*: the impersonal gift to the gods/*communitas* and the reciprocal gift to give, receive and return the gift to other members of *communitas*. I now extend these observations by proposing that the *munus* be theorized as a symbolic practice in which the members of *communitas* represent to themselves their being-in-common. Further, the collective representation of that to whom/what the *munus* is given constitutes and renews the political we. The reciprocal gift differs from the impersonal gift in being delinked from the political we; it is a space of connection and disconnection rather than being-in-common. Provisionally I call the space of the reciprocal gift the communal. The transcendental structure of *communitas* is thus characterized by a division between the political and the communal.

Roberto Esposito's theory of community firmly rejects the political dangers associated with affirming community as identity. Community occupies the space of the negative; its members have nothing in common, which is to say they are bound to the imperative of giving themselves away: a community of expenditure. Yet, as Nancy (23) implies in his preface to the French edition of Esposito's *Communitas*, community may be defined negatively as the absence of the proper and the proprietary subject, and yet still be characterized by the political "we." Nancy invites thinking about community as negative unity that possesses the qualities of place, sense and orientation. If the *munus* is an obligatory gift practice engaged in by community members (collective and individual), the political "we" of community may be conceptualized as a symbolic practice associated with the non-reciprocal impersonal gift.

The Durkheimian tradition has taken as its problematic the question of how collective social and political association is possible. In a Durkheimian sense, social collectivities have two levels of organization: interdependent practices and collective representations, the latter a symbolic practice through which members represent to themselves the associations of which they form part. Durkheim's classic discussion of the symbolic formation of the social took place in his commentary on the totem in *The Elementary Forms of the Religious Life*. He there interpreted the totem as a sacrificial practice in which social actors gather together on ritual occasions to kill and eat a particular animal or plant – the totem – in contrast to the prohibition in their everyday practice to refrain from hunting/gathering and eating that same animal or plant. In this sacrificial practice the totem signifies and constitutes the collective unity of the group; the ritual eating of the totem re/constitutes membership in the collectivity. In subsequent work by the École de sociologie, Durkheim's analysis of collective representations was extended from religion to the symbolic formation of the political (Richman). The Durkheimian tradition has had a complex relation of engagement and disengagement with scholarship that has, since the 1960s, inquired into the discursive/semiotic formation of the political we. This earlier Durkheimian work was augmented by Claude Lefort's conceptualization of the political as the symbolic constitution of a common fate made knowable only when it is represented. The symbolic foundation of the political establishes a "space that can, despite its internal heterogeneity, perceive itself in its entirety" (Lefort 3). These analytics of collective representations and the symbolic formation of the political do not presuppose the political as homogeneous.

In Esposito's theory of *communitas*, the *munus* is an intentional practice, a gift of and to. If *communitas* is constituted through the obligation to give, the donor(s) must give to. As donors are prohibited from being recipients of gifts, thereby suspending the reciprocities and social ties formed through the gift cycle of gift–reception–return, the only other to whom the gift might be given is to *communitas*/gods, that is, as I have argued above, the *munus* is governed by Mauss's fourth obligation: the obligation to give to the gods and their representatives. Following Durkheimian hermeneutic conventions, I extend "gods" to include symbolic representations used by members of a human multiplicity to perform their existence as a collectivity.[8] The *munus* is then a symbolic practice through which members form and renew the self-awareness of their collective existence – their being-in-common – in *communitas*.

The lack of a symbolic element in Esposito's philosophy weakens his treatment of the political and, by implication, *communitas*. The elision of collective representation arguably reduces *communitas* to social ties and networks formed through reciprocal gift relations as distinct from the impersonal gift to *communitas*. By way of historical example, in the Ancient Roman world the *munus* was a duty/gift to the city as an overarching political unit rather than a gift to kin, friends and businesses. The obligation to perform official duties on behalf of the city/*communitas* implies that the *munus* is linked to a political structure, membership in which binds the donor to make contributions to the collectivity. The absence of collective representations and the symbolic leaves untheorized the city and all it stands for at the level of the political. Severed from the political we, the gift dissolves into interrelations without being-in-common.

One of the hallmarks of Esposito's work on *communitas* consists of a refusal to theorize community as positive belonging for fear of its repeating a dangerous past and present of totalitarian mythologizing. There appears no necessary reason, however, to identify the political we as such with totalitarianism nor, as one can see from the example of the internally heterogeneous city, to postulate the political we as either whole or undifferentiated. The political may operate as a space of sharing and division.

This analysis suggests that the minimal transcendental structure of *communitas* has both vertical and horizontal vectors. The *munus* as the obligation to give without receiving introduces a vertical vector into community. This obligation is unidirectional and asymmetric, which indicates that status inequality obtains between *communitas* and its members. As demonstrated above, Esposito also conceptualizes the *munus* as horizontal exchange among community members that gives rise to reciprocity and mutuality. The members of community thus have a vertical relation to community and a horizontal relation to each other, with the former having primacy over the latter.

The vertical structure of *communitas* is evidenced in Esposito's account of Christian theology in *Communitas* (9–12). In Pauline theology, Esposito writes, the universe is understood as the gift of God to humanity. In the Christian *munus*, all souls are obliged to feel gratitude and make return gifts, although as created beings humans are incapable of returning a "true gift" (10) to repay the divine one. The Christian God becomes the other to whom humanity must donate, an example of what Mauss called the "obligation to give to the gods" (Mauss's fourth obligation), who, as Godelier observes, are unreliable and sometimes capricious members of gift relations. In Christian theology, humanity is the recipient of an incalculable gift, thus assuring its subordination to the God/donor in a debt that may in principle never be repaid. The Christian *munus* thus constructs *communitas* in its verticality. The vertical structure of *communitas* is repeated in the historical case of the Ancient Roman world where public *munera* consisted of obligations incumbent on citizens and residents to give back to the city from whose membership they regarded themselves as having benefited; the public *munus* is a gift to the *city*. In each of these cases, theological and historical, the gift is donated to a collective representation that symbolizes *communitas* in its negative unity. A debt then arises to those/that which has given the members of *communitas* their bodies and protected their existence. Yet the impersonal gift to the god/city/*communitas* cannot occur without god/city/*communitas*' being signified.

In other passages Esposito conceptualizes the *munus* as creating relations of mutual indebtedness, exchange, reciprocity and even mutuality. Reciprocal relations presuppose another form of transcendental gift relation other than the vertical one produced by the unilateral obligation to give. The horizontal axis, in contrast, constructs reciprocal indebtedness as community members have the obligation of giving to the other and all are, over time, givers and receivers. The horizontal relations are congruent with status inequalities and finite indebtedness: a space of being-in-difference rather than being-in-common. The vertical axis paradoxically places the donor rather than the recipient in the position of permanent status inequality and infinite indebtedness to the gods/city/*communitas*. The vertical and horizontal vectors of the transcendental gift system functional antithetically, the vertical being the principle of the political and the horizontal the principle of the communal.

The distinction between the vertical, impersonal gift and the horizontal, reciprocal gift gives rise to a number of preliminary observations. First, the impersonal gift introduces a power relation into *communitas* between those that represent the gods/city/*communitas* and those that do not. The space of *communitas* is thus asymmetrically expropriative. This bears mention as Esposito's political philosophy addresses the question of power at best indirectly. The vertical, political vector of the transcendental gift system could be conceptualized not as voluntary expenditure of self but as compulsory expropriation that provokes resistance. It might be responded that Esposito considers conflict and violence as intrinsic to *communitas*, and thus resistance may fall within the transcendental *communitas* as well. However, where expropriation is theorized as the gift of care for the other, questions of domination and imperial order may be occluded.[9] Second, from these remarks on power it follows that the abstract insertion of the political into *communitas* should lead to further inquiries into what forms of the political may be compatible with an expropriated model of *communitas* where the earth, bodies, labor, things and actions are not perceived as property. Third, the political may be instituted through the gift rather than sacrifice. The significance of this statement awaits fuller thinking elsewhere.

One final note: the proposal to consider *communitas* as comprised of two distinct gift relations, the impersonal and the reciprocal, rests on the assumption that the *munus* may be identified as a gift. The concept of the *munus* tests the expansive limits of the gift. To the extent that Esposito models his discussion of the *munus* and community on Ancient Roman history, it is not apparent why legally obligatory taxes and labor services should be called a gift relation in a philosophical or social scientific sense, although *dono/donum* was used in Latin with reference to *munera*. Given its absence of apparent voluntariness, the public *munus* lacks the double structure of the gift. More generally, not all political and social obligation has the structure of the gift, even where actors desire or consciously agree with their obligations. The contemporary payment of personal income tax is, for instance, a legal obligation, an expropriation, but it is not a gift precisely because it is a legal obligation. As obligatory transfer to *communitas*, the *munus* is a task and duty of corporate membership.

Esposito's philosophy of *communitas* theorizes the political in relation to the gift. Working from Derrida's intervention in gift theory, he turns the philosophy of the gift away from its main focus during the twentieth century: the relation between the gift and the economic. In this article I have sought to contribute to the splendid project of the political gift by tracing Esposito's debt to Mauss in his conceptualization of *munus* as the law of community. I have asked Maussian questions that have given rise to Maussian conundrums about to whom/what the gift is given and how reciprocities arise under the conditions reduced to the single obligation to give. I have treated the *munus* as an instance of Mauss's fourth obligation, and interpreted it broadly to mean the obligation to give to the gods/the political and their representatives. Such a gift necessarily presupposes the symbolic formation of the gods/*communitas* as a representation of the being-in-common of *communitas*. Stated otherwise, the *munus* is a gift to someone/something that is the collective representation of *communitas* and the symbolic formation of the political we. The *munus* may be termed an impersonal gift in contrast to the social ties and interrelations that are formed through the reciprocal gift: communality without the common. The transcendental *communitas* thus has a double structure given through the non-reciprocal impersonal gift and the reciprocal gift. The impersonal gift in its verticality links the members of *communitas* to the political we, while the reciprocal gift in its horizontality connects members to each other by forming ties and differentiations. The double structure of *communitas* includes vertical, impersonal gifts to the political we and its representatives and horizontal, reciprocal gifts among community members, the former giving rise to being-in-common and the latter to being-in-difference. Further reflection is called for concerning which forms of the political we are compatible with *communitas* stripped of homogeneity and the metaphysics of the proper.

notes

Thanks enormously to the editors of this special issue of *Angelaki*, Greg Bird and Jon Short, for their generous and incisive comments on my draft. I am grateful for María del Rosario Acosta's interest in my work and will continue to think through her comments. I am indebted to Frank Pearce for ongoing conversations on the Durkheimian tradition. My colleague Hira Singh kindly alerted me to debates among historians of India about the status of land grants as gifts.

1 See the home page of *La Revue du MAUSS*, available <http://www.revuedumauss.com/> (accessed 5 Apr. 2013); my translation.

2 There is a fourth obligation in Mauss's *The Gift* (14–17) that has been much less discussed than the first three: the obligation to give to the gods. In this paper I later argue that the *munus* is an instance of the fourth obligation.

3 Although those in the dominant cultures of the Global North primarily associate gifts with dyadic transactions between individuals, gifts may involve vast networks of linked exchange, the classic example of which would be the kula rings of the Western Pacific that stretch thousands of miles. Gifts are given between states, kin groups and users of transnationalized digital communications; the reduction of the gift exchange to interpersonal dyads forms a fundamental misunderstanding of gift relations.

4 The norm of reciprocity has been contested for decades. Anthropologists currently accept that certain types of gift are not bound by the obligation to return. See the work of Godbout on gifts to strangers and Giesler on digital file sharing.

5 The "Introduction" to *Communitas* is subtitled "Nothing in Common."

6 The failure to conceptualize the empirical object of study would be an elementary methodological error for any Durkheimian, and very surprising for Mauss given his long collaboration with Durkheim (Fournier). I set aside here the question of whether Derrida's claim is accurate.

7 The distance between the gift and exchange has widened in anthropology with recent interest in gifts to strangers (Godbout) and gifts that may only be kept (Weiner).

8 The collective representation of the political we may be desacralized while those of the gods are always sacralized. Where Durkheimians in the first half of the twentieth century identified the political with the sacred, contemporary work in that tradition explores the divergence of the political from the sacred, partly with the aim of aligning the political with a space of debate. Thanks to Frank Pearce for his observations on this topic.

9 The limits of the gift for theorizing political obligation have been debated among historians. Contemporary historians of India have been concerned about whether land grants made by Indian kings during the feudal period should be conceptualized as gifts or as feudal grants. See Thapar; Sharma.

bibliography

Bird, Greg. "Roberto Esposito's Deontological Communal Contract." *Angelaki* 18.3 (2013): 33–48. Print.

Corbier, Mireille. "Munus, Munera." *Brill's New Pauly* n.d. Web. 7 May 2013. <http://referenceworks.brillonline.com.ezproxy.library.yorku.ca/browse/brill-s-new-pauly>.

Derrida, Jacques. *Given Time: I. Counterfeit Money.* Trans. Peggy Kamuf. Chicago and London: U of Chicago P, 1992. Print.

Diprose, Rosalyn. *The Bodies of Women: Ethics, Embodiment and Sexual Difference.* London and New York: Routledge, 1994. Print.

Esposito, Roberto. *Bíos: Biopolitics and Philosophy.* Trans. Timothy C. Campbell. Minneapolis: U of Minnesota P, 2008. Print.

Esposito, Roberto. *Communitas: The Origin and Destiny of Community.* Trans. Timothy C. Campbell. Stanford: Stanford UP, 2010. Print.

Esposito, Roberto. "Community, Immunity, Biopolitics." Trans. Zakiya Hanafi. *Angelaki* 18.3 (2013): 83–90. Print.

Esposito, Roberto. "Donner la technique." *La Revue du MAUSS* 2.6 (1995): 190–206. Print.

Esposito, Roberto. *Immunitas: The Protection and Negation of Life.* Trans. Zakiya Hanafi. Cambridge: Polity, 2011. Print.

Esposito, Roberto. *Terms of the Political: Community, Immunity, Biopolitics.* 2008. Trans. Rhiannon Noel Welch. New York: Fordham UP, 2013. Print.

Fournier, Marcel. *Marcel Mauss.* Paris: Fayard, 1994. Print.

Giesler, Marcus. "Consumer Gift Systems." *Journal of Consumer Research* 33 (2006): 283–90. Print.

Godbout, Jacques T., with Alain, Caillé. *The World of the Gift.* Trans. Donald Winkler. Montreal and Kingston: McGill-Queen's UP. Print.

Godelier, Maurice. *The Enigma of the Gift.* Trans. Nora Scott. Chicago: U of Chicago P, 1999. Print.

Jenkins, Tim. "Derrida's Reading of Mauss." *Marcel Mauss: A Centenary Tribute.* Ed. Wendy James and N.J. Allen. New York and Oxford: Berghahn, 1998. 83–96. Print.

Lefort, Claude. *The Political Forms of Modern Society.* 1986. Boston: MIT P, 1988. Print.

Mauss, Marcel. *The Gift: The Form and Reason for Exchange in Archaic Societies.* Trans. W.D. Halls. New York and London: Norton, 1990. Print.

Mauss, Marcel. "Gift-gift." 1924. *Œuvres.* Paris: Minuit, 1969. 3: 46–51. Print.

Nancy, Jean-Luc. "Conloquium." 1999. *Minnesota Review* 75.1 (2010): 101–08. Print.

Richman, Michele H. *Sacred Revolutions: Durkheim and the Collège de Sociologie.* Minneapolis: U of Minnesota P, 2002. Print.

Sharma, Ram Sharan. *Rethinking India's Past.* Delhi and New York: Oxford UP, 2009. Print.

Thapar, Romila. *Cultural Pasts: Essays in Early India History.* New Delhi and New York: Oxford UP, 2000. Print.

Weiner, Annette B. *Inalienable Possessions: The Paradox of Keeping-While-Giving.* Berkeley and Los Angeles: U of California P, 1992. Print.

Index

For Product Safety Concerns and Information please contact our EU representative GPSR@taylorandfrancis.com Taylor & Francis Verlag GmbH, Kaufingerstraße 24, 80331 München, Germany

Batch number: 10726294

Printed by Printforce, the Netherlands